D1093136

The Handbook of Organic and Fair Trade Food Marketing

The Handbook of Organic and Fair Trade Food Marketing

Edited by
Simon Wright and Diane McCrea

Blackwell
Publishing

© 2007 by Blackwell Publishing Ltd

Blackwell Publishing editorial offices:
Blackwell Publishing Ltd, 9600 Garsington Road, Oxford OX4 2DQ, UK
Tel: +44 (0)1865 776868
Blackwell Publishing Professional, 2121 State Avenue, Ames, Iowa 50014-8300, USA
Tel: +1 515 292 0140
Blackwell Publishing Asia Pty Ltd, 550 Swanston Street, Carlton, Victoria 3053, Australia
Tel: +61 (0)3 8359 1011

First published 2007 by Blackwell Publishing Ltd

ISBN: 978-1-4051-5058-3

1005274283

Library of Congress Cataloging-in-Publication Data

Handbook of organic and fair trade marketing / edited by Simon Wright and Diane McCrea.
p. cm.
"Useful organic and fair trade websites": p.   .
Includes bibliographical references and index.
ISBN-13: 978-1-4051-5058-3 (hardback : alk. paper)
1-4051-5058-0 (hardback : alk. paper) 1. Natural foods industry--Handbooks, manuals, etc. 2. Natural foods--Marketing--Handbooks, manuals, etc. 3. Farm produce--Marketing--Handbooks, manuals, etc. 4. Beverages--Marketing--Handbooks, manuals, etc. I. Wright, Simon, 1957- II. McCrea, Diane.
HD9000.65.H36 2007
641.3'020688--dc22
2006025809

A catalogue record for this title is available from the British Library

Set in 10/13pt Sabon
by Sparks, Oxford – www.sparks.co.uk
Printed and bound in Singapore
by Markono Print Media Pte Ltd

For further information on Blackwell Publishing, visit our website:
www.blackwellfood.co.uk

Contents

Foreword

Sainsbury's is the leading organic and fair trade retailer within the UK and we are proud to be sponsoring this handbook. No other sectors within the UK food and drink industry are growing at the rates of organic and fair trade, and no other sectors are experiencing such fundamental shifts in customer buying patterns and attitudes. Organic and fair trade are leaving their niche status behind. The last few years have seen attitudes change dramatically and as a result many different customer groups are buying organic and fair trade for many different reasons.

The need to react to changing customer demands and the increasingly universal appeal of organics was fundamental to the re-launch of our organic range as Sainsbury's SO organic in September 2005. The plans put in place for the re-launch, with over 100 new lines and lower prices on 100 everyday items, have been fundamental in making organics more accessible to many of our customers, encouraging those previously not buying organic food and drink to *try something new*.

Fair trade's success mirrors organics. Never before have customers been so engaged on issues of poverty alleviation and never before have so many realised how much difference fair trade makes. We were the first major UK retailer to sell fair trade products. Recently, we have built on this heritage by converting all of our own-brand rose bouquets to fair trade and placing the largest ever single order of fair trade cotton for our Sport Relief T-shirts. These initiatives, coupled with a steady stream of additions to both the branded and own-label ranges, make us the largest fair trade retailer in the UK.

The rapid journey that has brought organics and fair trade to the position that they enjoy today is a theme that runs through many of the chapters in this book. Many of the contributing manufacturers, retailers and brands chart their history from small or family-run businesses to much larger companies, operating successfully in multinational markets. These impressive accounts bear testament to the speed at which the organic and fair trade markets have grown, but also reflect the ability of these companies to quickly adapt to and satisfy rapidly changing customer demands.

Readers of this book can access at first hand the insights and strategies that the authors have adopted and the challenges that they have overcome in order to drive forward their products within the organic and fair trade markets.

If you are new to this sector, you will find the *Handbook* useful. Even if you are established in the sector, you will still find much of interest. I hope it motivates you to do more in this important and rapidly developing sector.

Justin King
Group Chief Executive, Sainsbury's

Contributors

Ruth Bailey, Sainsbury's Supermarkets Ltd, 33 Holborn, London EC1N 2HT, UK

Lorraine Brehme, Patley Wood Farmhouse, Silkhay, Netherbury, Bridport, Dorset DT6 5NG, UK

John Bowes, High House Farm, Ings, Nr Windermere, Cumbria LA23 1JR, UK

Martin Cottingham, 12 Long Meadow, Bristol, BS16 1DY, UK

Paola Cremonini, Cremonini Consulting, Via Dagnini 27, 40137 Bologna, Italy

David Croft, 6 Roundbarn, Blackburn Road, Turton, Bolton, Lancashire BL7 0QA, UK

Ella Heeks, Abel & Cole Ltd, Waterside Way, Wimbledon, London SW17 0HB, UK

Graham Keating, Yeo Valley Farms (Production) Ltd, The Mendip Centre, Rhodyate Blagdon, Nr Bristol BS40 7YE, UK

Harriet Lamb, Fairtrade Foundation, Room 204, 16 Baldwin's Gardens, London EC1N 7RJ, UK

Elaine Lipson, New Hope Natural Media, 1401 Pearl Street, Suite 200 Boulder, CO 80302, USA

Diane McCrea, 127 Havannah Street, Cardiff Bay, Cardiff CF10 5SF, UK

Petra Mihaljevich, Duchy Originals, The Old Ryde House, 393 Richmond Road East, Twickenham TW1 2EF, UK

Hubert Rottner, Nagelhof 1, 91174 Spalt, Germany

Amarjit Sahota, Organic Monitor, 79 Western Road, London W5 5DT, UK

Craig Sams, 106 High Street, Hastings, East Sussex TN34 3ES, UK

Elisabeth Winkler, *Living Earth*, The Soil Association, South Plaza, Marlborough Street, Bristol BS1 3NX, UK

Simon Wright, O&F Consulting, The Old Bakery, 8A Replingham Road, London SW18 5LS, UK

Chapter 1

The International Market for Organic and Fair Trade Food and Drink

Amarjit Sahota
Director, Organic Monitor

Introduction

This chapter gives an overview of the global market for organic and fair trade products. The focus is on the organic food and beverages market as its market size dwarfs that of the fair trade products market.

This chapter is divided into six sections. An overview of the global market for organic food and drink is given followed by separate sections for western Europe, North America, Asia and Australasia. These regions have the most important markets for organic products in the world. Other regions such as Africa, Latin America, and central and eastern Europe are becoming important although they have low significance compared to the regions covered. A section on the global market for fair trade products is followed by future projections in the conclusion.

Organic global overview

The global market for organic food and drink was valued at $US27.8bn in 2004. Global sales surpassed the $US30bn mark in 2005 with the highest growth occurring in North America. Organic food and drink sales in the USA and Canada are expanding by over $US1.5bn a year. Although organic farming is practised throughout the world, the most important markets are in North America and Europe, which comprise 96% of global revenues, as shown in Figure 1.1.

The proportions of organic farmland are more evenly split across the globe, as shown in Table 1.1. About 31.5 million hectares of farmland was certified organic in 2005.[1] Australasia leads with 12.2 million hectares followed by Latin America (6.4 million hectares) and Europe (6.3 million hectares). Organic land area does not equate to organic food production since not all the land is farmed – for example, countries such as Australia and Argentina have vast organic land areas that are not used for farming.

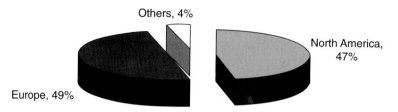

Figure 1.1 Distribution of global organic food revenues, 2005. Note: all figures are rounded. Source: Organic Monitor.

Table 1.1 Breakdown of organic farmland by region, 2005.

	Organic farmland (m ha)	% of total
Africa	1.026	32.6
Asia	4.064	12.9
Europe	6.500	20.6
Latin America	6.363	20.2
North America	1.378	4.4
Oceania	12.171	38.6
TOTAL	31.503	100.0

Note: All figures are rounded. Source: FiBL Survey 2005/2006

Table 1.2 lists the countries with largest areas of organic farmland. Australia dominates because of significant areas of the great outback used by organic cattle farmers. Argentina also has large areas of organic pasture, whereas China has seen a large rise in organic farmland in recent years. The country is fast becoming a global supplier of organic ingredients.

Important consumer countries with large areas of organic farmland are Italy, USA, Germany and the UK. Countries such as China, Brazil and Uruguay are important producers of organic crops but the majority of production is for export markets.

Alpine countries have the largest proportion of organic farmland relative to total land area, as shown in Table 1.3. Over 10% of the farmland in Lichtenstein, Austria and Switzerland is managed organically.

Table 1.4 lists the countries with the most organic farms. Mexico leads with 120000 organic farms. The country has many small producers of organic crops such as coffee, cocoa, avocados and papayas. Indonesia has many small producers of organic crops such as vegetables, coffee and herbs. Most of the 36639 organic farms in Italy are in the south of the country.

Table 1.2   Countries with largest areas of organic farmland, 2004.

	Organic farmland (×1000 ha)
Australia	12 127
China	3 467
Argentina	2 800
Italy	954
USA	889
Brazil	888
Germany	768
Uruguay	759
Spain	733
UK	690
Chile	640
France	534
Canada	489
Bolivia	364
Austria	345
Mexico	295
Czech Republic	260
Peru	260
Greece	250
Others	3 993
TOTAL	31 502

Note: All figures are rounded. Source: FiBL Survey 2005/2006

Table 1.3   Countries with largest proportion of organic farmland, 2005.

Country	Organic area as % of total agricultural area
Liechtenstein	26.4
Austria	13.5
Switzerland	11.3
Finland	7.3
Sweden	6.8
Italy	6.2
Czech Republic	6.1
Denmark	5.8
Portugal	5.4
Estonia	5.2
Uruguay	5.1
Slovenia	4.6
Germany	4.5
UK	4.4

Note: All figures are rounded. Source: FiBL Survey 2005/2006

Table 1.4   Countries with largest number of organic farms, 2005.

	Number of organic farms
Mexico	120 000
Indonesia	45 000
Italy	36 639
The Philippines	34 990
Uganda	33 900
Tanzania	30 000
Kenya	30 000
South Korea	28 951
Peru	23 400
Others	239 902
TOTAL	622 782

Note: All figures are rounded. Source: FiBL Survey 2005/2006

Organic Europe

Organic market overview

The European market for organic food and drink was the largest in the world until it was overtaken by North America in 2005. The European market was worth $US13.7bn in 2004 and sales are estimated to have reached $US14.4bn in 2005. Sales of organic products increased by about 5% in 2005 although some countries reported higher growth rates.[2]

Germany has the largest market for organic foods in Europe, valued at about $US4.5bn in 2005. Sales are growing by 10–12% a year as the number of channels offering organic products expands. A growing number of conventional supermarkets are offering organic products and the number of organic supermarkets continues to increase with 40 new organic supermarkets opening in 2004 alone.

The UK market continues to show healthy growth with 10% growth reported in 2004. The UK market, valued at about $US2bn, is the third largest in the world. Like most European countries, the highest growth is being observed in the fruit and vegetable, and meat and dairy categories. Much of the growth in the UK market is occurring in non-supermarket channels such as organic food shops, box schemes and farmers' markets. A growing number of catering and food service companies in the UK are also offering organic foods.

The Italian and French markets are the next most important in Europe but growth rates have slowed in these countries. Other important markets are in Switzerland, Austria, Sweden and the Netherlands.

There is a small market for organic foods in central and eastern Europe (CEE) with the region comprising less than 3% of European revenues. Although the amount of organic farmland in CEE countries is rising, mostly primary products such as grains, seeds and herbs are grown. There is a lack of organic food processing in the region with a high volume of the organic crops exported to western Europe. Organic food companies export some of the resultant finished goods back to the region. Demand for organic products is growing rapidly in countries such as the Czech Republic and Hungary, especially in the country capitals.

Since western Europe accounts for most revenues in the European organic food industry the focus is on this region in the sections that follow.

Organic farmland

Europe has 6.5 million hectares of organic farmland of which about 5.3 million hectares are in western Europe.[1] Central and eastern European countries with large areas of organic farmland are the Czech Republic (260 120 hectares), Ukraine (241 980 hectares) and Hungary (128 690 hectares).

Organic farmland comprises 3.3% of the total farmland in western Europe and the amount of organic farmland in the leading countries is given in Table 1.5. All the countries are members of the European Union (EU) except Switzerland.

Italy has the largest amount of organic farmland, 954 361 hectares, representing 18% of the EU15 total. The country has the fourth largest amount of organic farmland in the world. About 36 639 farms practice organic agriculture in Italy, the highest number in western Europe. Most of the organic farms are in the south with

Table 1.5   The amount of organic farmland in western European countries, 2005.

	Organic farmland (×1000 ha)	% of total
Austria	328.8	12.90
Belgium	24.2	1.73
Denmark	165.1	6.20
Finland	160.0	7.22
France	550.0	1.86
Germany	734.0	4.30
Greece	244.5	6.24
Ireland	28.5	0.65
Italy	1 052.0	6.86
Luxembourg	3.0	2.00
The Netherlands	41.9	2.17
Norway	38.2	3.68
Portugal	120.7	3.17
Spain	725.3	2.84
Sweden	207.5	6.80
Switzerland	110.0	10.27
UK	695.6	4.42
EU 15	5 080.5	3.17

Note: All figures are rounded. Source: FiBL Survey 2005/2006

Sicily and Sardinia accounting for over half of Italian organic farmland. The amount of organic farmland has decreased from 1.23 million hectares in 2001 because many farmers quit the profession.

Germany, with 767 891 hectares, has the second largest amount of organic farmland in western Europe. Organic farmland represents 4.5% of total farmland in the country. About half of the 16 603 organic farms are in the southern states of Baden-Württemberg and Bavaria.

Spain has 733 182 hectares of organic farmland, representing less than 3% of total farmland in the country. Much of the organic crops are exported to other European countries with the country having a small market for organic food and drink.

The amount of organic farmland in the UK has stabilised after years of high growth. Over half the 690 270 hectares of organic farmland is in Scotland where there are large areas of pastures. 4.4% total farmland is organic.

France has 534 037 hectares of organic farmland making up 1.8% of total farmland. The amount of organic farmland has stabilised in line with consumer demand for organic products.

Other countries with high levels of organic farmland are Austria, Greece, Portugal and Sweden. Greece has shown a large rise in organic farmland partly because many livestock farmers have opted for organic production methods. The country had just 31 100 hectares of organic farmland in 2001.

Sweden has the largest area of organic farmland in the Nordic region. Organic farmland in Denmark and Finland also comprises more than 5% of total farmland.

The share of organic farmland in Austria at 13.5% is the second highest in the world. Organic farmland has represented over 10% of Austrian farmland since 2000. Switzerland also has a large share of organic farmland, 11.3%. Large areas of alpine pastures in these countries are managed organically.

One reason behind Austria, Finland and Sweden having large areas of organic farmland is these countries were late entrants to the EU, joining in the mid 1990s. Their late entry allowed most farmers to practise extensive forms of agriculture and some of these farmers converted to organic farming methods upon joining the EU.

There is much variation in the amount of organic farmland between European countries. Two factors are broadly responsible for the variation:

(1) Financial subsidies/conversion grants given to farmers
(2) The number of extensive farms

The European agricultural sector is heavily subsidised by the EU in the form of the Common Agricultural Policy. Financial incentives are arguably more important to organic farmers since they need to be encouraged to practise extensive forms of agriculture. Farmers also need to be compensated for the conversion period, which can be up to three years, during which they cannot sell organic products. There is a link between the level of financial subsidies provided to farmers and the amount of organic farmland. Italy has the highest land area of organic farmland partly because of the high level of subsidies provided to organic farmers. There was only a high take-up rate of organic farming in the UK after a new government increased financial support to British organic farmers in the late 1990s.

The number of extensive farms in a country also has a bearing on the conversion rate to organic farming. Extensive farms have low reliance on chemical inputs such as pesticides and the conversion process is easier than for larger intensive farms that are more dependent on chemical inputs. This factor is partly responsible for the large number of organic farms in Austria, Switzerland, Finland and Sweden.

In the UK, there was a large rise in organic farmland in the late 1990s because many extensive farms in Scotland converted to organic agriculture. The large amount of organic farmland in Germany was also partly a result of many extensive farms converting to organic farming. East Germany had many state-owned extensive farms and some converted to organic farming after re-unification. More farmers converted to organic agriculture in eastern Germany in the five years after re-unification then had done so in western Germany in the preceding 70 years. The same trend is currently being observed in EU accession countries, which are showing large rises in organic farmland.

Organic sales channels

Figure 1.2 shows the sales breakdown of organic foods by marketing channel.[2] Mainstream retailers make up about 50% of organic product sales in Europe. The highest market share is in Scandinavian and Alpine countries where over 85% of sales are from supermarkets. In countries such as Germany, the Netherlands and France, mainstream retailers usually focus on core products such as organic milk, potatoes, bakery products and cereals. A much wider product assortment is found in specialist retailers.

Specialist retailers accounted for most organic food sales until the late 1990s when their market share was overtaken. Organic food shops and other specialists accounted for about 46% of total sales in 2005. The market share has been in decline because of increasing sales from the supermarkets. Only in two leading countries – France

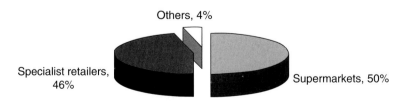

Figure 1.2 Organic food sales breakdown by channels, 2004. Note: all figures are rounded. Source: Organic Monitor.

and Germany – do specialist retailers still have the majority of organic food sales. Specialist retailers were also the most important sales channels in the Netherlands and Italy up to 2002.

The other marketing category includes direct marketing, professional box schemes and the catering and food service (CFS) sector. About 4% of organic food sales went to these channels in 2005. There is increasing demand from the CFS sector with a growing number of restaurants, bars and cafés serving organic food. The public sector is also an important buyer of organic fresh produce with many national governments promoting organic food consumption in hospitals, schools and government buildings.

Although supermarkets account for most sales at the total level, their market share is in decline in some European countries. In the UK, for example, the supermarket share shrunk in 2003 and 2004 because of a higher volume going into other sales channels. Direct marketing and specialist retailers are showing a higher increase in sales than the supermarkets, possibly because British consumers preferring to buy regionally produced organic products.

Organic supply chain overview

The largest number of organic food and beverage companies is in Europe. There are estimated to be over 8000 companies involved in organic food production and distribution in Europe. Table 1.6 lists some of the most important companies in the European organic food industry.

Table 1.6 lists companies that focus on organic foods and related products.[3] The leading traders and wholesalers of organic foods are not listed. The list focuses on some of the leading suppliers of organic foods.

Many conventional food companies are involved in organic foods. Indeed, many sectors of the organic food industry are dominated by non-organic food companies:

- The organic meat products market is controlled by large conventional meat companies. For instance, most of the organic meats in the UK are supplied by large slaughterhouses such as St Merryn and Anglo Beef processing. Large conventional meat companies such as Danish Crown, Dumeco and Swedish Meats have strong market positions across Europe.
- Large European dairies have a high market share in the organic dairy products market. Arla Foods, Europe's largest dairy group, is the dominant supplier of organic milk in Denmark, Sweden and the UK. Other dairy companies with a high market share include Dairy Crest (UK), Lactalis (France), Campina Melkunie (the Netherlands) and Valio (Finland).

Most conventional food companies have come into the organic food industry by starting production lines of organic products. Some have entered by acquiring dedicated organic food companies. Important acquisitions in 2005 include:

- Green & Black's, the leading organic chocolate brand in Europe, being bought by Cadbury Schweppes in May 2005
- The Dutch bank ABN Amro buying Nutrition et Sante, a leading French organic soya products company, in November 2005
- Premier Foods acquiring Cauldron Foods, a leading British organic vegetarian foods company, in October 2005

Table 1.6   Selected list of leading organic food companies in Europe, 2006.

Company	Country	Company Details
Hipp	Germany	Largest organic baby food manufacturer in the world
Andechser Molkerei	Germany	Largest organic dairy company in Germany
De Vau Ge	Germany	A leading organic health and organic food company
Biotropic	Germany	A leading German organic fresh produce company
Rapunzel	Germany	Leading German brand of organic foods
Pro Natura	France	A leading supplier of organic fruit & vegetables
Triballat Noyal	France	Large producer of organic dairy & non-dairy beverages
Bodin et Fils	France	Europe's largest producer of organic poultry
Distriborg	France	Leading organic food company in France
Organic Farm Foods	UK	Leading organic fresh produce company in the UK
Yeo Valley Organic	UK	One of Europe's leading organic yogurt producer
Duchy Originals	UK	A leading British brand of organic foods
Bioitalia	Italy	A leading Italian organic food company
Probios	Italy	Leading Italian brand of organic foods
Alpro	Belgium	Europe's leading producer of organic dairy alternatives
Hain Celestial	Belgium	Owns Natumi, Lima Foods, Biomarche and other firms
Royal Wassenen	The Netherlands	Well-diversified Dutch conglomerate
EOSTA	The Netherlands	Largest supplier of organic fresh produce in Europe

Source: Organic Monitor

Organic retailer overview

Specialist retailers

There are up to 10 000 specialist retailers in Europe.[2] About half are organic food shops and the number continues to expand as investment increases and as chains develop. Some of the important developments since 2002 have been:

* Whole Foods Market, the largest natural and organic food supermarket in the world, entering Europe by acquiring Fresh & Wild (UK) in February 2004.
* Rewe, the leading conventional supermarket in Germany, is investing in a chain of dedicated organic supermarkets with the first opening in Düsseldorf in May 2005.
* The number of organic supermarkets in Germany is expanding fast with 40 new stores opening in 2004. Leading chains such as Basic and Denn's Bio opened new stores in 2005.

Table 1.7 lists the most important organic food shop chains in Europe. Germany and France have the most organic food retail chains. Southern European countries tend to have more dedicated organic food retailers than northern European countries.

Alnatura is the leading chain in Germany with 16 organic supermarkets in 2004. The average size of an Alnatura supermarket is 500 m^2 housing about 6000 organic products. The retailer reported €117m sales in 2004 and there are plans to open five more supermarkets in the short term. Alnatura is also a wholesaler of organic products, selling to German retailers as well as retailers in Central Europe. Basic AG is the second leading organic supermarket chain in Germany. The retailer reported a 39% sales increase to €39m in 2004 with one new store opening in Hamburg that year. Basic AG is planning to open five new supermarkets in 2005. Dennree, the leading wholesaler of organic products in Germany, has entered the organic retail sector with its Denn's Bio retail chain.

Fresh & Wild is the leading chain of organic food shops in the UK. The seven-store chain was acquired by Austin-based Whole Foods Market for €57m in 2004. All seven stores are based in southern England with six in London. Whole Foods Market plans to open its first American-style supermarket in London in January 2007. Planet Organic is the second leading British chain with three stores.

Collobora B'io is the largest chain of organic food shops in Italy. There are approximately 230 franchised stores with the B'io brand in the country. The franchise system has been set up by the organic food wholesaler, Ecor. NaturaSí is a chain of 38 organic supermarkets, located in the north and centre of the country. NaturaSí has also set up franchised butchers and restaurants. Bottegae Natura is a chain of organic food shops owned by the Ki Group wholesaler. Terra d'Incanto is another organic food shop chain set up by the wholesaler, Baule Volante.

Table 1.7 Leading organic food shop chains in western Europe, 2005.

Supermarket	Country	No. of shops	Details
Alnatura	Germany	16	Retailer and wholesaler of organic products
Basic AG	Germany	11	Large supermarket format stores
Denn's Bio	Germany	7	Subsidiary of the leading wholesaler Dennree
SuperBiomarkt AG	Germany	8	Some stores located in cathedrals
Supernatural AG	Germany	5	Based in North-Rhine / Westphalia region
Erdkorn	Germany	5	Stores are located in Northern Germany
Eat Organic (e.o.)	Germany	5	Berlin-based organic supermarket
Fresh & Wild	UK	7	Owned by US-based Whole Foods Market
Planet Organic	UK	3	London-based organic food retailers
Collabora B'io	Italy	230	Franchised stores set up by Ecor wholesaler
NaturaSi	Italy	38	Chain of organic supermarkets in major cities
Emporio Alcatraz	Italy	10	Franchised chain of organic food shops
Bottegae Natura	Italy	9	Owned by the Ki Group wholesaler
Terra d'Incanto	Italy	7	Set up by Baule Volante wholesaler
Biocoop	France	232	Fast growing organic food retailer
La Vie Claire	France	122	Long established natural food chain
Biomonde	France	33	Stores located in south of France
Naturalia	France	25	Paris-based organic food shop chain
Satoriz	France	19	Shops located in Eastern France
La Vie Saine	France	10	Grouping of 10 organic food shops
L'Eau Vive	France	7	Grenoble-based retail chain
Terra Verda	Spain	15	Mainly in the Valencia area
Ecoveritas	Spain	10	Organic supermarkets mainly in Barcelona

Source: Organic Monitor.

In France, Biocoop with 232 stores has the largest number of organic food retailers in Europe. The typical store size is over 200 m² with over 7000 organic and natural products. Biocoop reported about €166m sales in 2004 and it plans to increase its number of stores to 300 in 2006.

La Vie Claire is the oldest natural food shop chain in Europe, established in 1946. It operates a franchise of 122 stores in France. Biomonde is a chain of 30 organic food shops in the southwest and southeast of France; it plans to expand to 40 stores in 2005. Naturalia is a chain of organic food supermarkets and it also operates an online home delivery scheme in the capital. Other important French chains are Satoriz, La Vie Saine and L'Eau Vive.

Ecoveritas has ambitious expansion plans to become the dominant organic food retailer in Spain. There are plans to increase the 10 Véritas organic supermarkets to 40 in the coming years. Half of the current number is in Barcelona. Terra Verde has 15 organic food shops, mainly in the Valencia area.

Mainstream retailers

Mainstream retailers are responsible for the majority of organic food sales in Europe, however, most multiple retailers have only a basic assortment of organic lines. The product range is increasing in Germany, France and Spain.

Table 1.8 lists the leading supermarkets marketing organic foods in western European countries. The approximate number of organic items in their product range is given as well as the names of their private labels.

Tegut is one of the most successful conventional supermarkets marketing organic foods in Europe. The retailer has 400 supermarkets in the states of Hessen and Thüringen, which market up to 1800 organic products. Organic foods sales, at €110m, make up about 10% of Tegut's total sales. Rewe is the largest food retailer in Germany with about 4000 supermarkets, although organic foods comprise a small fraction of total sales. Organic food sales in Edeka supermarket account for about 2% of its total sales.

Tesco, the second largest retailer in Europe, has the largest market share for organic food sales in the UK. Organic food sales were estimated to have surpassed €450m in 2004. Sainsbury's is the second leading retailer in terms of organic food revenues with roughly €420m sales in 2004. Waitrose, a much smaller retailer with 140 supermarkets, has the highest market share in terms of total food sales; organic products comprise 15% of all food sales.

Co-op Italia is the second leading food retailer in Italy but it has the highest market share for organic food sales. Many of its organic products are marketed under its Bio-logici Co-op private label. The market share of Esselunga is in decline because it significantly reduced its range of organic products in 2003 and 2004.

Carrefour is the dominant retailer of organic foods in France and organic products are found in all its hypermarkets. About 25% of its organic products are marketed under its Carrefour Bio private label. Auchan is the second most important French retailer of organic products with about 700 items in its product range.

Albert Heijn is the dominant food retailer in the Netherlands and it accounts for most organic food sales. Although organic food sales continue to increase, the number

Table 1.8 List of major supermarkets marketing organic foods in western Europe, 2005.

Supermarket	Country	Organic products	Private labels
Tegut	Germany	1800	Tegut Bio
Edeka	Germany	500	Bio-Wertkost, Bio-Gutfleisch
Rewe	Germany	300	Füllhorn
Tesco	UK	1200	Tesco Organic
Sainsbury's	UK	1200	Sainsbury's Organic
Waitrose	UK	1200	Waitrose Organic
Coop Italia	Italy	350	Bio-logici Coop
Esselunga	Italy	200	Esselunga Bio
Carrefour	France	800	Carrefour Bio
Auchan	France	700	N/A
Albert Heijn	The Netherlands	275	AH Biologisch
Delhaize	Belgium	600	Bio
Billa	Austria	700	Ja, Naturlich!
Coop Schweiz	Switzerland	1200	Naturaplan
SuperBrugsen (FDB)	Denmark	800	Økologisk Natura
Gröna Konsum (KF)	Sweden	800	Änglamark
Kesko	Finland	500	Pirkka
NKL	Norway	400	N/A
El Corte Inglés	Spain	300	N/A

Source: Organic Monitor.

of organic items has not changed for a number of years. Delhaize is the dominant food retailer in Belgium and organic foods make up 3% of all its food sales.

Billa supermarkets account for most organic food sales in Austria. The Ja Naturlich! private label has 400 organic products and generated €200m sales in 2004. Spar is the second leading retail chain with its Natur*Pur private label.

Coop Schweiz is the leading organic food retailer in Switzerland. The co-op reported that organic food sales had reached €406m in 2004. A high portion of sales are from its Naturaplan private label. Migros is the second leading supermarket for organic foods. The two retailers account for about 75% of organic food sales in the country.

SuperBrugsen and Gröna Konsum are the leading retailers of organic foods in Denmark and Sweden, respectively. The number of organic products shrunk in these retailers to roughly 800 items in 2004 because of reorganisation of the Coop Norden group, which owns these supermarkets. NKL, also part of the Coop Norden group, is the leading retailer of organic foods in Norway. Kesko, the leading wholesale-based retailer in Finland has the largest market share for organic foods in the country.

El Corte Inglés accounts for most retail sales of organic foods in Spain although Carrefour is showing a large rise in market share since it launched its private label for organic products in 2003.

Organic North America

Organic market overview

The North American market for organic products is the fastest growing in the world. Organic food and drink sales are increasing by about $US1.5bn a year. With the market value estimated at $US14.5 bn, the USA has the largest market for organic foods in the world. The market was worth about $US12.2bn in 2004. Although organic fresh produce and dairy products comprise most revenues, all sectors are reporting high growth rates. American manufacturers have launched organic versions of many popular products. For instance, organic versions of American favourites such as pizza, hot dogs and peanut butter are now available in retailers.

Increasing distribution is a major driver of market growth. The traditional retail outlets for organic foods were natural food shops such as Whole Foods Market and Trader Joe's; however, mainstream grocery retailers now comprise the majority of organic food sales. The range of organic products continues to expand in super-markets such as Safeway, Albertson's and Kroger. Even Wal-Mart, renowned for its low-cost products, has jumped on the organic bandwagon by announcing that it wants to increase market share.[4]

The Canadian market is also reporting high market growth. Sales of organic food and drink were estimated to have reached $US900m in 2004. As in the USA, a major driver is increasing distribution with a growing number of supermarkets launching organic products.

Demand for organic products is so high in the USA and Canada that many industry sectors are experiencing supply shortages. Producers are importing organic products from across the globe because of insufficient production in North America. For instance, organic seeds and grains are coming in from Europe and Asia; organic herbs and spices from Latin America and Asia; organic beef is imported from Australasia and Latin America. Supply shortages have led Stonyfield Farms, the largest organic yogurt producer, to look at importing organic milk powder from New Zealand.

A feature of the organic food industry in North America is that large food companies dominate almost every sector. Multinationals such as Dean Foods, General

Mills and Pepsi-Cola have strong market positions. Specialist organic food companies are being acquired by large food companies, private equity firms or organic food conglomerates such as Hain Celestial and SunOpta.

North America is the only region in the world in which organic food companies are listed on the stock exchange. Organic food companies such as Hain Celestial and SunOpta as well as retailers such as Whole Foods Market and Planet Organic are publicly listed.

Organic farmland

There are about 1.4 million hectares of certified organic farmland in North America. The breakdown of organic farmland by countries is shown in Table 1.9.[1]

The USA with 889 048 hectares (2 196 874 acres) has the fifth largest amount of certified organic farmland in the world. According to the USDA's Economic Research Service (ERS), the amount of organic farmland increased from 378 527 hectares (935 450 acres) in 1992 to 544 852 hectares (1.35 million acres) in 1997. Organic farmland expanded by 63% between 1997 and 2003 to approximately 0.89 million hectares (2.2 million acres).

Many American states observed a decline in organic farmland in 2002 and 2003 because the USDA implemented the National Organic Programme (NOP). The largest declines were reported in southern states such as Georgia, Louisiana, South Carolina, Tennessee and West Virginia with the main reason being that certification was typically provided by small non-profit organizations that dropped certification services when the NOP was introduced. These states are expected to show a rise in organic farmland in the coming years as new certifiers emerge and existing ones expand their services to these states.

Although there has been a large expansion in organic agriculture since the early 1990s, organic farmland accounted for just 0.2% of total farmland in 2003. Organic pasture represented just 0.1% of total pasture, and organic cropland comprised 0.4% of total cropland. Specialised crops such as lettuce, carrots and apples, had the largest market share, between 1% and 5%.

Organic farming is practised in every American state except Mississippi. In 2003, Texas had the most acreage with 313 248 acres and California was second with 220 953 acres followed by Alaska with 174 647 acres and North Dakota with 147 780 acres.

Table 1.9 Organic farmland in North American countries, 2004.

	Organic farmland (ha)	% of total
Canada	488 752	0.72
USA	889 048	0.22
TOTAL	1 377 800	0.29

Note: All figures are rounded. Source: FiBL Survey 2005/2006

The USA had 8035 certified organic farms in 2003. California leads with almost 2000 producers. Wisconsin is second with 659 organic farmers, Washington has 541, Iowa 448 and Minnesota 392. States with many certified organic operators are typically those with a high portion of small organic farms. California dominates because of a high concentration of organic fruit and vegetable growers.

Cropland comprises 1.45 million acres of certified organic farmland in the USA. California is the leading state with 176 507 acres followed by North Dakota with 128 963 acres. Most of the organic cropland in California is used for fruit and vegetable production. Organic farmland in North Dakota is used to grow organic wheat, soybeans and other field crops. Other important states for organic crop production are Minnesota, Colorado, Idaho, Iowa, Montana, South Dakota, Texas and Wisconsin.

Over 3000 facilities were processing and distributing certified organic products in 2005. Most companies are concentrated on the Pacific Coast and have a 41% share of the total. California has the largest number, nearly 800. In comparison, over half the states had less than 30 facilities.

Canada had 485 288 hectares (1 199 172 acres) of organic farmland in 2004. Although the country is ranked fourteenth in the global table of organic farmland, the acreage represents merely 0.72% of Canadian farmland with 3670 organic farms, representing 1.50% of all farms.

Saskatchewan is the leading province for organic agriculture with 1245 farmers that manage 292 297 hectares. The province has 60% of Canada's organic farmland and 34% of organic farms. Other leading provinces are Alberta (95 375 hectares), Ontario (30 777 hectares) and Quebec (27 919 hectares). Ontario has many organic food processors and traders, making up 22% of the country's total, Quebec is an important producer of organic fruit, vegetables, greenhouse crops, field crops and maple products.

Organic field crops are mostly grown in Canada and important provinces for these crops are Saskatchewan, Ontario, Manitoba and Alberta. Organic grain and oilseed producers farm 194 201 hectares (479 678 acres) of organic farmland; the most important crops are wheat, oats and flax. Canada is one of the leading exporters of organic grains in the world with half of total exports going to the European Union (EU), 41% to the USA and the rest to Asia.

Organic supply chain overview

There are estimated to be over 1000 companies involved in the production and supply of organic foods in North America. The vast majority is in the USA, which has many organic food growers, processors and distributors. Table 1.10 lists some of the most important companies in the North American organic food industry.[3]

Hain Celestial is the world's largest natural and organic food company, reporting $US544m sales in 2004. The company has grown by a series of acquisitions of

Table 1.10   List of selected organic food companies in North America, 2006.

Company	Country	Company details
Earthbound Farms	USA	Specialises in organic fruits and vegetables
General Mills	USA	Owns several dedicated organic companies
Hain Celestial	USA	The world's largest natural and organic food company
Whitewave Foods	USA	Owned by Dean Foods
Stonyfield Farm	USA	Largest organic yogurt producer in the world
Amy's Kitchen	USA	Leading producer of organic ready-meals and soups
Organic Valley	USA	Leading co-operative of organic farmers
Smucker Quality Beverages	USA	Owns several organic beverage brands
Cal-Organics	USA	A leading supplier of organic fresh produce
Golden Temple	USA	A leading supplier of organic cereals and tea
Nature's Path Foods	Canada	The leading producer of organic cereals
SunOpta	Canada	Diversified company with range of organic products

Source: Organic Monitor.

American and European natural products and organic food companies. It is a leader in over 13 of the top natural and organic food categories in the US market.

General Mills is one of American's leading food manufacturers. The company produces organic flour and organic sugar and it has acquired a number of dedicated organic food brands. Its subsidiary, Small Planet Foods, owns the Cascadian Farm and Muir Glen brands. General Mills also owns SunRise, which packs organic breakfast cereals.

WhiteWave Foods was created by the merger of Horizon Organic and WhiteWave. The two companies were bought by Dean Foods in 2003 and merged into a single entity a year later. Horizon Organic is the leading brand of organic dairy products while Whitewave operates in the soya drinks market with its Silk brand.

Organic Valley is the leading farmers' co-operative in North America with 723 organic farms in 22 states. The Organic Valley brand is one of the most prominent in the organic food industry, found on organic dairy products, fresh produce, eggs, juices and processed foods. Although the co-operative produces around 110 organic products, it is the second largest organic dairy processor in the USA.

Earthbound Farms is the largest supplier of organic fresh produce in North America. Based in California, it supplies organic fruits, vegetables and salads across the USA and it also exports to Europe and Asia.

Stonyfield Farms is the world's leading producer of organic yogurts. The company was bought by the French food group Danone in 2004.

Smucker Quality Beverages is one of the largest beverage companies in the USA. It owns several organic brands such as Santa Cruz Organic and R.W. Knudsen. Smucker Quality is owned by The J.M. Smucker Company.

Nature's Path is the leading producer of organic cereals in North America. The Canadian company also produces organic breads, waffles and cereal bars. Its organic products can be found all over the USA and Canada and it also exports to Europe and Asia.

SunOpta is the largest Canadian organic food company. It has grown by acquiring several natural and organic food companies in North America.

Organic sales channels

Figure 1.3 shows the rough breakdown of organic food sales in North America by sales channels. It is estimated that about 60% of organic foods are bought from conventional grocery channels. Mainstream retailers overtook specialist retailers as the dominant channel for organic foods in the late 1990s.[4]

Conventional grocery channels refer to supermarkets, discount and club stores, mass merchandisers and convenience shops. The market share is expanding as retail penetration of organic food and drinks increases.

In the USA and Canada, all the leading supermarkets are selling organic products. The product range varies between retailers; however, some of the most popular products are organic fresh produce, bakery products, dairy products and cereals.

Specialist retailers were responsible for the majority of organic food sales. The sector includes about 25 000 natural food retailers, organic and health food shops and similar retailers. Most of these retailers are in big cities such as New York, San Francisco, Los Angeles and Toronto. Large supermarket-store formats such as Whole Foods Market and Trader Joe's deliver the majority of organic food sales.

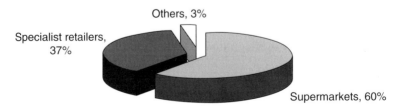

Figure 1.3   Organic Food Sales Breakdown by Channels, 2004. Note: all figures are rounded. Source: Organic Monitor.

Organic retailer overview

Mainstream retailers are playing an increasingly influential role in the North American organic food industry. Indeed, competition between retailers is expected to drive market growth in the coming years.[4]

Conventional grocery retailers are attracted by the high market growth rates in the organic food industry. Some see marketing organic products as a way of raising revenues, others a way to combat competition from other retailers. The growing threat from Whole Foods Market has prompted supermarkets such as Safeway to adopt an organic retailing strategy. Important developments in conventional grocery channels include:

- Wal-Mart announcing in March 2006 that it is to step up marketing activities for organic foods and double its product range in the coming months.
- Supervalu opening Sunflower Market, a dedicated organic food supermarket, in January 2006. The retailer plans to have a chain of 50 Sunflower Market supermarkets in the USA.
- Publix Super Markets planning to open two dedicated organic food supermarkets under the GreenWise name in 2006.
- Stop'n Shop experimenting with organic food store-in-store concepts in its supermarkets. The holistic centres are operated in conjunction with Wild Oats.
- Safeway re-launching its private label for organic foods under the O Organics banner in December 2005. The launch was supported by media advertising.

These retailer initiatives are making organic foods more widely available to North American consumers. They are also building confidence in the industry, encouraging farmers to grow more organic crops and processors to make more organic products. Indeed, a major talking point in 2006 was how Wal-Mart's organic strategy would impact on the industry considering the undersupply situation.

Table 1.11 lists the leading retailers marketing organic foods in the USA and Canada. All are mainstream retailers, except Whole Foods Market, Trader Joe's, Wild Oats and Planet Organic.

Kroger is the largest food retail group in the USA with over 3700 stores. Although Wal-Mart has overtaken Kroger in terms of grocery sales, Kroger remains the largest dedicated grocery chain in the country.

Kroger launched its 'Naturally Preferred' private label for organic and natural products in April 2003. Its private label has more than 275 organic and natural products in its 2500 supermarkets in 2005. The product range includes organic baby food, pasta, cereals, snacks, milk, eggs, juices and soya products.

Albertson's is the second leading supermarket chain in the US with more than 2500 stores. Organic foods have been in Albertson's supermarkets since the late 1990s and over two-thirds of its stores were marketing organic products in 2005. The widest range of organic food and beverages are in its southern Californian

Table 1.11   List of leading retailers marketing organic foods in North America, 2006.

Supermarket	Country	Details
Ahold USA	USA	Launched Nature's Promise label for organic foods in 2004
Albertson's	USA	Organic products marketed under Essensia private label
Kroger	USA	Organic foods marketed under Nature's Prefered label
Safeway	USA	Re-launched private label under O Organics name in 2005
Trader Joe's	USA	Second leading American natural food retailer with 250 stores
Wild Oats	USA	Operates 114 natural food stores in USA and Canada
Whole Foods Market	USA	World's leading natural food retailer has 175 stores in 3 countries
Loblaws	Canada	Largest Canadian retailer of organic foods with PC Organics label
Planet Organic	Canada	Leading chain of natural food shops in Canada

Source: Organic Monitor.

supermarkets, which include fruit and vegetables, dairy products, dairy alternatives, meat alternatives, baby food, beverages, snacks and frozen foods. A growing number of organic products are marketed under its 'Essensia' private label, which was launched in spring 2003.

Safeway is one of the largest food retailers in North America. Its stores have a wide range of organic products that include fruit and vegetables, dairy products, snacks, cereals and beverages. The supermarket relaunched its private label under the 'O Organics' name in December 2005. The private label had 150 organic products in spring 2006 that included cereals, bakery products, dairy products, snacks and frozen foods. O Organics is found in all its 1775 outlets.

Ahold USA initially launched organic foods in its Giant-Carlisle stores in the New York–Long Island area. The company launched a private label for organic and natural foods, 'Nature's Promise' in its Giant and Stop and Shop stores in September 2004. About 100 organic products were initially launched under the private label with the number increasing to 200 in 2005. The product range includes organic milk, butter, eggs, juices, as well as natural cookies, broth and chips.

Wal-Mart, the world's largest retailer with over 4800 stores in three continents, is poised to become a leading retailer of organic foods. The retailer was marketing organic foods in some of its American supercentres in 2005. In March 2006, Wal-Mart announced that it would double its organic product selection in the coming months. The move is part of the retailer's strategy to expand its customer base and compete more strongly with other retailers. It plans to have 400 organic products in its supercentres and Neighbourhood Markets.

Whole Foods Market was founded in 1980 as one small store in Austin, Texas. It has grown to become the world's largest natural food supermarket chain with 175 stores in three countries. All its supermarkets are in the USA except three in Canada and seven in the UK. Its entry in the UK was by the acquisition of the organic food shop chain Fresh & Wild in February 2004. The company is listed on the Nasdaq stock exchange and was listed 30th in the Fortune 100 Best Companies to Work For 2005.

Whole Foods Market stores average 32 000 square feet and they have a wide range of organic products that include fruit and vegetables, dairy and meat products, beverages, and non-food items such as personal care products and clothing. Organic products account for about 40% of Whole Foods Market sales. The retailer markets more than 1300 items under six lines of private labels. All its supermarkets in the USA are certified organic by Quality Assurance International.

Trader Joe's has grown to become the second largest chain of natural food shops in the USA. The retailer started as a convenience store chain called 'Pronto Markets' in the Los Angeles area in 1958. It has expanded to have 250 stores in 20 states, mostly on the west and east coasts. Trader Joe's remains privately owned.

Trader Joe's markets upscale grocery products such as health foods, organic products and nutrition supplements. It focuses on developing its private labels (such as Trader Joe's, Trader Jose's and Trader Giotta's), which have over 2000 products and account for about 80% of its sales. The average size of a Trader Joe's store is 10 000 sq feet. Its stores have a comprehensive range of organic products although few are marketed under manufacturer brand names.

Wild Oats is the third largest natural food supermarket chain in North America. It was founded in 1987 in Bolder, Colorado and currently operates 110 stores in 24 American states and four in Canada. Its retailers operate under the Capers Community Market, Henry's Farmers Markets, Sun Harvest Farms, and the Wild Oats Natural Marketplace names.

Wild Oats has a comprehensive range of organic products, which make up about 40% of its total sales. As well as organic products, the retailer markets natural products, nutrition supplements and environmentally friendly household products in its stores. Wild Oats has been experiencing difficulties since 2000. Its strategy of building big-store concepts has not been popular with shoppers, and it has since reversed its strategy by focusing on developing small stores driven by service.

Loblaws is Canada's largest supermarket chain with over 1000 stores. The supermarket launched 20 items in its President's Choice Organics (PC Organics) range in June 2001 and the number had risen to over 300 in 2005. PC Organics products are found in all Loblaws retailers and the range includes fresh fruit and vegetables, bread and flour, baby food, cheese and dairy products, frozen foods, pasta and condiments. The retailer stated that sales of organic foods were expanding by about 30% per annum. High growth is partly because of stores widening their range of organic products. The retailer has set a target that organic foods should represent 5% of all food sales in its stores.

Planet Organic is the leading natural food supermarket chain in Canada. The first store opened in Edmonton in June 2001 and it was operating five stores in 2005. Planet Organic has made a number of acquisitions in recent years as part of its bold plans to consolidate the fragmented natural and organic products industry in Canada. It bought a chain of 35 Sangster's Health Centres retailers in 2004. The same year, it acquired Trophic Canada, a manufacturer and distributor of vitamins, minerals and herbal supplements. The expansion spree of Planet Organic continued in 2005 with the acquisition of 17 Newfound Health stores.

Planet Organic supermarkets have in-store bakeries, a deli counter and a wide range of organic products that include fresh produce, salads, dairy products, processed foods and desserts. Non-food items include natural personal care products and textiles.

Organic Asia

Organic market overview

Although home to about 60% of the world population, Asia has a small market for organic products. Production of organic crops is increasing in the Asian region; however, sales are largely in the most affluent countries. The Asian market for organic foods was valued at about $US750m in 2004.[3]

The largest market is in Japan, which is estimated at $US400m. This market size only refers to sales of organic products meeting Japanese Agricultural Standards (JAS). Sales are much higher if other organic products are included, especially those directly marketed by producers to consumers. Organic products such as fruit, vegetables, rice and green tea are mostly grown in Japan. Processed foods are imported from countries such as Australia, USA and Germany.

The highest market growth is occurring in China, which has the largest area of organic farmland in Asia. Production of organic crops has increased significantly in recent years. The growing affluence of Chinese consumers and expanding expatriate community is developing a domestic market for organic foods. Foreign supermarkets such as Carrefour are importing organic foods from Europe to meet growing demand. The number of organic food shops, especially in the major cities, is increasing to meet burgeoning consumer demand.

Other Asian countries with large markets for organic foods are South Korea, Taiwan, Singapore and Malaysia. Asian consumers in these countries have relatively high disposable incomes and organic products are popular as they are perceived to be healthier and more nutritious than non-organic foods. Food scares such as bird flu are making Asian consumers more aware of the production differences between organic foods and non-organic foods. There is also growing demand for organic foods in countries such as India, Thailand and the Philippines. Unlike the first-tier countries, much of the demand for organic foods is met by domestic products in these countries.

Health food shops are the most important sales channels for organic foods in Asian countries. Organic foods are mostly sold in the major cities such as Tokyo, Seoul, Singapore, Hong Kong and Taipei.

Organic farmland

The Asian continent has 4.06 million hectares of certified organic farmland, comprising 13% of the global total. Table 1.12 lists the countries with the largest areas of organic farmland. [1]

China dominates in terms of organic farmland with roughly 3.5 million hectares under organic cultivation. The country has experienced a large rise in organic farmland in recent years as sustainable forms of agriculture become popular. Farmers are interested in organic farming practices as they realise the damage caused by chemical pesticides and synthetic fertilisers. Much of the growth, however, has been export driven with the Chinese government encouraging organic farming because of the economic benefits.

Bangladesh has the second largest area of organic farmland with 177 700 hectares. Other important countries are India (114 037 hectares), Indonesia (52 882 hectares), Kazakhstan (36 882 hectares) and Russia (30 000 hectares).

The Asian region is expected to continue to show a large rise in organic farmland in the coming years. A growing number of governments are introducing national organic programmes, which are encouraging farmers to convert to organic practices. Government initiatives are expected to drive growth in India, Thailand, Malaysia and the Philippines.

Table 1.12 Leading Asian countries with organic farmland, 2004.

Country	Organic farmland (ha)	% of total
Azerbaijan	20 105	0.43
Bangladesh	177 700	1.97
China	3 466 570	0.60
India	114 037	0.06
Indonesia	52 882	0.12
Japan	29 151	0.56
Kazakhstan	36 882	0.02
South Korea	28 218	1.46
Pakistan	20 310	0.07
Russia	30 000	0.01
Others	88 144	–
TOTAL	4 063 999	0.24

Note: All figures are rounded. Source: FiBL Survey 2005–2006.

Organic Australasia

Organic market overview

The Australasian continent comprises almost 40% of the world's organic farmland; however, the market size is a fraction of the global total. Sales of organic food and drink were estimated at about $US250m in 2004.[3]

Production of organic foods in Australia and New Zealand has traditionally been export orientated. Much of the organic crops grown in these countries are sold in different parts of the world. For instance, organic lamb from New Zealand is sold in western Europe and North America; organic beef from Australia is marketed in Japan and the USA. Other important exports include organic seafood, apples, honey and wool.

The domestic market for organic products in New Zealand and Australia is growing at a steady rate. Most sales are of organic fresh products such as fruit, vegetables, milk and beef, although there is also an increase in organic food processing activity. The number of conventional food retailers selling organic products is increasing, and new organic food shops continue to open.

Most sales of organic foods were initially from health food shops and specialist retailers; however, mainstream retailers now account for the majority of organic food sales. The leading supermarkets in Australia and New Zealand are marketing organic foods, albeit a small product range. Indeed, large conventional food companies such as Heinz Wattie's Australasia, Fonterra and Zespri are becoming increasingly active in producing organic food and beverages.

Organic farmland

There are 12.172 million hectares of organically managed land in Australasia. The breakdown of organic farmland by country is shown in Table 1.13.[1]

Almost all organic farmland is in Australia and the vast majority of this is used as pasture by cattle farmers. This organic farmland is not comparable with that in Europe and North America where large amounts of organic farmland are used for horticulture and arable farming. Hence, although Australia has the largest amount

Table 1.13   The Amount of organic farmland in Australasian countries, 2005.

Country	Organic farmland (×1000 ha)	% of total
Australia	12 126.6	2.71
Fiji	0.2	0.04
New Zealand	45.0	0.26
Total	12 171.8	2.62

Note: All figures are rounded. Source: FiBL Survey 2005–2006.

of organic farmland in the world, the level of organic food production is a fraction of that in countries with much less organic farmland.

New Zealand has 45 000 hectares of organic farmland, the second largest in the continent. Unlike Australia, most of this is used to produce organic fruit, vegetables, meat and dairy products. There is a small amount of organic farmland in Fiji.

Global market for fair trade products

The fair trade products market is showing a higher percentage growth than the organic food and drink market, although market revenues are a fraction of the organic food market. The global market for fair trade products was valued at about $US900m in 2005. As for organic foods, demand is concentrated in the most affluent regions of the world.[5]

The market for fair trade products, although much smaller than the organic products market, is growing at a rapid rate. Countries such as the UK are reporting market growth rates in excess of 20% per annum. The global market for fair trade products was valued at about $US1bn in 2004.

Like the organic food market, demand is concentrated in the most affluent countries of the world. Table 1.14 gives the market size of fair trade products of the leading countries. The three leading country markets – USA, UK and Switzerland – comprise two-thirds of global revenues.

Table 1.14   Leading markets for fair trade products, 2004.

Country	Estimated retail sales ($US million)
USA	259.7
UK	248.7
Switzerland	164.6
France	84.3
Germany	69.6
The Netherlands	42.4
Italy	30.3
Canada	21.2
Austria	19.1
Belgium	16.5
Denmark	15.7
Sweden	6.7
Ireland	6.1
Norway	5.8
Japan	3.0
Luxemburg	2.4
Finland	9.1
Total	1005.1

Note: All figures are rounded. Source: Fairtrade Labelling Organizations International.

Fair trade products are produced in over 32 countries by more than 600 000 producers. Most are located in Latin American countries such as Costa Rica, Brazil and Guatemala, and African countries such as Uganda, Kenya and Cameroon. Production is also increasing in Asian countries such as Indonesia, India, Thailand and China.

A major difference between organic and fair trade products is that the latter is almost entirely produced in southern countries. There is thus a large disparity between countries that produce and those that consume fair trade products.

Table 1.15 lists the leading fair trade products in the world market. Food and drink products, especially commodities, are the dominant fair trade products.

Coffee is the best-known fair trade product. The USA is the largest consumer of fair trade coffee in the world, consuming 3574 metric tonnes in 2003. The Netherlands, the UK, Germany, France and Switzerland are the other important markets for fair trade coffee, having sales of over 1000 metric tonnes a year.

Almost 2000 tonnes of fair trade tea was sold in 2003. The UK has the largest market for fair trade tea, consuming over 1000 tonnes a year.

Banana is the leading fair trade fruit with over 80 000 tonnes sold in 2004. Many producers supply bananas which are both fair trade and organic, a combination becoming increasingly popular in the UK and the Netherlands.

Apart from bananas, over 5000 tonnes of other fair trade fresh fruit is sold a year. Pineapples, oranges and mangoes are the major products. The market is showing high growth as it is starting from a very low base. In line with the rise in processing of fair trade processing, juice production is increasing with over 4500 tonnes sold in 2004.

Table 1.15   Leading fair trade products, 2004.

Product	Volume	2003/04 growth (%)
Coffee (tonnes)	24 222	26
Tea (tonnes)	1 965	29
Bananas (tonnes)	80 641	58
Fresh Fruit (tonnes)	5 175	299
Cocoa (tonnes)	4 201	56
Sugar (tonnes)	1 960	173
Honey (tonnes)	1 239	6
Juices (tonnes)	4 542	107
Rice (tonnes)	1 384	154
Dried Fruit (tonnes)	238	934
Beer (litres)	62 934	1631
Wine (litres)	617 744	–
Sport-balls (items)	56 479	-66
Flowers (stems)	13 008 850	–
Ice tea (litres)	532	233
Other (tonnes)	46	32

Note: All figures are rounded. Source: Fairtrade Labelling Organizations International.

Fair trade wine is produced mainly by South African growers. Over half a million litres were sold in the first year of sales with the UK the main market. Fair trade beer is showing a large rise in sales, 1631% in 2004.

Cocoa, sugar and rice are important fair trade commodities. The UK is the largest buyer of fair trade cocoa with sales of 903 tonnes in 2004. Switzerland is the largest consumer of fair trade sugar and rice.

Germany and Switzerland are the leading markets for fair trade honey. Other important fair trade products are nuts and spices.

The fair trade sports ball was the brainchild of Rättvisemark, a Swedish organisation. It initiated fair trade sports ball production in Pakistan in 2001. Although sales fell from 165 125 in 2003 to 56 478 sports balls in 2005, a sharp recovery is expected in 2006 due to the World Cup.

Fair trade flowers were introduced for the first time in 2004. Over 13 million stems were sold in the first year. Kenya is an important producer of fair trade roses and the UK is a leading market.

Conclusions

Consumer demand for organic and fair trade products is growing across the world. With sales expanding by about $US2.5bn a year, the value of the global market exceeded $US31bn in 2005.

Demand for organic and fair trade products is concentrated in North America and Europe, which comprise about 97% of global revenues. The rest of the world accounts for a mere 3% of global revenues, and most are generated from Japan and Australia.

Why is consumer demand concentrated in the most affluent countries of the world? Two factors are believed to be responsible.

First, the price premium of organic and fair trade products restricts demand to countries where consumers have high purchasing power. This is a factor why most sales are in countries where there is a sizeable middle-class population. This is also a reason why large cities represent the majority of organic food sales in Asian countries.

Second, education or awareness of organic and fair trade products is important. As consumers become more educated and informed, they are more inclined to buy organic and fair trade products because of factors such as concern for the environment, ethical and societal considerations or health reasons.

The organic food market has the most revenue because it is more established. Some predict that fair trade products could follow the growth of the organic food and drink industry and become a multi-billion dollar industry. Although it is growing by over 20% a year, the fair trade market is likely to remain a fraction of the size of the organic food market unless there are some structural changes.

Organic crops are also grown in every continent whereas fair trade production is mainly in Latin American and African countries. It has been suggested that the manufacturing and producing base for fair trade products should broaden. A wider range of products need to have the fair trade logo if they are to gain greater consumer acceptance. Fair trade production needs to move away from commodities and more into processed foods and value-added products, otherwise consumer demand will be restricted by a narrow product range.

References

1 Willer, H. and Yussefi, M. (2006) *The World of Organic Agriculture: Statistics and Emerging Trends*. IFOAM & Research Institute of Organic Agriculture (FiBL). Frick, Switzerland.
2 Organic Monitor (2005) *The European Market for Organic Fruit & Vegetables*. Organic Monitor, London.
3 Organic Monitor (2006) *The Global Market for Organic Food & Drink*. Organic Monitor, London.
4 Organic Monitor (2006) *The North American Market for Organic Meat Products*. Organic Monitor, London.
5 Fair Trade Labelling Organizations International (2005) *Annual Report 2004/2005*. Fair Trade Labelling Organizations International, Bonn.

Chapter 2

The Organic Consumer

Martin Cottingham
Freelance Writer/Communications Consultant and former Marketing Director,
the Soil Association

Elisabeth Winkler
Editor, *Living Earth*

Introduction

Until 20 years ago organic farming was hardly heard of or talked about. Organic food was produced by a dedicated coterie of farmers for a lucky few to enjoy. From the mid 1980s onwards, however, there were breakthroughs in both awareness and availability that were to become the building blocks for one of the most powerful ethical consumer movements of the twenty-first century.

Public awareness of food safety and environmental issues began to increase significantly in the 1980s as a succession of highly publicised food scares undermined trust in the mass-produced and often anonymous offerings of the mainstream food industry. At the same time, organic food started to become available to a wider public as producers persuaded a wider network of health-food stores and then supermarkets to stock organic produce.

The market established by the determination of those pioneering producers gradually developed a media and consumer-driven momentum of its own, attracting a growing mass of support from shoppers. Today, over three-quarters of UK households buy some organic food, and the sales value of organic products is close to £4 million per day.

This chapter examines the make-up and motivations of the increasingly discerning and demanding consumers who are fuelling this explosive growth. Much of the research upon which it is based was conducted on behalf of the Soil Association, the leading organisation promoting and certifying organic food and farming in the UK and a driving force in the growth of the market. Each year the Soil Association's *Organic Market Report* (formerly published as *The Organic Food and Farming Report*) provides a comprehensive overview of the UK organic market, including a section focusing on consumer attitudes and motivations.

A broadening appeal

A recent Soil Association report shows that more people and a wider cross-section of society are buying organic food than ever before. A MORI poll in 1999[2] found that only a third of people were organic shoppers, and buyers tended to be predominantly female (56%), middle-aged and from the better-off ABC1 social strata. By 2006, a zOmnibus poll conducted for the Soil Association by Market Tools Inc[1] found that the proportion of consumers knowingly buying organic food in the previous six months had risen to 63% – nearly two out of three; 39% were buying organic food at least once a month and 22% once a week or more (see Figure 2.1).

The 2006 poll showed that organic food was still more popular with women (69% were buyers) than with men (63%) and with those over 25 (66% were buyers) compared to those under 25 (59%). Also over half those in the most disadvantaged social brackets – C2, D and E – identified themselves as organic shoppers. Purchasing was even at 57% in the lowest income bracket included in the research – those earning under £16000 a year. At least 56% of consumers in every age bracket and at least 58% in every region and nation of the UK were found to be buying organic products.

A mass of motivations

Organic consumers have a plethora of motivations, reflecting the wide range of benefits associated with organic food and farming. From a marketing perspective this can be both a blessing and a curse. On the one hand, organic products have a broad appeal because the various health, environmental, animal welfare and social

Base: Total respondents	1000	
MORE THAN ONCE A WEEK (NET)	**101**	**10%**
Every day	10	1%
4–5 times a week	23	2%
2–3 times a week	68	7%
ONCE A WEEK TO EVERY 3 WEEKS (NET)	**237**	**24%**
Once a week	125	12%
Once every two to three weeks	112	11%
ONCE A MONTH OR LESS (NET)	**291**	**29%**
Once a month	51	5%
Two or three times in the past six months	128	13%
Once in the past six months	113	11%

Figure 2.1   A zOmnibus poll conducted on behalf of the Soil Association in 2006 found that the proportion of consumers knowingly buying organic food in the previous six months had reached 63%, with 22% buying once a week or more.

boxes they tick mean that they have the potential to attract many different segments of society. On the other hand, there are so many benefits that it can be difficult to come up with clear, compelling marketing messages.

In the government's organic action plan for England, for example, there is a declaration that organic farming 'is better for wildlife, causes lower pollution from sprays, produces less carbon dioxide, generates fewer dangerous wastes, operates to high animal welfare standards and increases jobs in the countryside.'[3] That's six benefits and six messages, before any of the health arguments is even mentioned. Once you add in higher vitamin and mineral content, reducing dietary exposure to pesticides, avoiding genetically modified (GM) crops and ingredients, avoiding antibiotic residues and minimising potentially harmful food additives, the number of benefits reaches double figures – and then there's food quality and taste to consider.

Despite the complexity, the messages are getting through. The same zOmnibus poll conducted in 2006[1] found that there is now strong acceptance of the benefits of organic food and farming among the general population, not just organic consumers. Of all the consumers surveyed, 49% (rising to 75% among regular organic consumers) believed that an organic diet was healthier; more than four times as many as disagreed with this view. Around three-quarters (and 88% of regular organic consumers) believed that organic production was kinder to the environment and wildlife, with only 3% disagreeing. Only 9% of consumers disagreed with the view that 'animal welfare is of higher quality on organic farms', whereas 52% (and 72% of regular organic shoppers) were convinced of the animal welfare benefits.

The relative importance of all these motivations in the UK was illuminatingly examined in the most extensive piece of consumer research the Soil Association has commissioned in recent years, conducted by Taylor Nelson Sofres (TNS) in 2003 and supported by the South West Regional and Welsh Development Agencies and Organic Centre Wales. Its key findings are summarised in *The Organic Food and Farming Report 2003* and described in more detail in a report aimed at Welsh organic businesses and published by the Welsh Development Agency.[4]

What was distinctive about this research was the extent to which it studied both consumer attitudes and buying patterns in a single project, mapping the two areas together to develop a rounded analysis of consumer motivations and buying behaviour.[4] Buying behaviour and associated attitudes were explored using spending data and questionnaire findings from the TNS Superpanel, which was made up of 15 000 households each issued with palm-top screening devices to scan every purchase they made. TNS also carried out an omnibus poll of 4000 adults. The aim was to establish the motivational triggers that persuade people to buy organic food for the first time, and what motivates them to 'trade up' to buy a wider range and/or to increase their overall spending. The result was a detailed segmentation of light, medium and heavy buyers, with a useful analysis of how their motivations and attitudes differ and change over time.

Understanding what drives heavy consumers is particularly important, as the organic market relies significantly on a relatively small core of households who are the biggest spenders. In 2003, TNS found that just 23% of buyers accounted for 84% of total spend.[4] In 2005, Market Tools found that only 9% of organic shoppers estimated that organic products accounted for over 50% of their food and drink spend, while 45% put their organic spending at under 10% (see Figure 2.2).[5]

TNS found that heavy organic consumers tended to be older and more upmarket than the average UK shopper. Over two-thirds of them were in social classes A, B or C1, compared to under 50% in the population at large. They were likely to have fewer children and were most heavily concentrated in London and south-east England. They came across as a discerning and self-confident group of consumers, far less price conscious than the average and much more likely to regard themselves as connoisseurs of food and wine, to seek out eco-friendly products and to read ingredient labels before buying. Over 80% of their spending on organic food and drink was in Tesco, Sainsbury's or Waitrose stores.

Taste and health – the main purchasing triggers

Taste and food safety concerns (avoidance of health 'negatives' such as pesticide residues rather than embracing 'positives' such as higher vitamin content) were found by TNS to be the main factors persuading people to try organic food for the first time. There was also a direct correlation between the extent to which consumers believed in the health and taste benefits of organic food and the number of categories (e.g. fresh produce, dairy, meat) into which they bought.

The research found, however, that light consumers only became serious organic shoppers when they were persuaded of the positive health, environmental and animal welfare benefits of eating organic. Taste remained an important factor in influencing increased consumption, but the more committed consumers were also characterised by a more sophisticated understanding of a wide range of benefits.

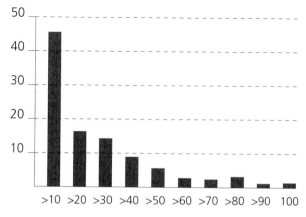

Figure 2.2   The percentage response to the question, 'What proportion of your food shop is organic?'

The implications of this for the marketing of organic food and drink are that compelling health messages and good-quality, tasty products are the best way to get new consumers on board, but public education about a wider range of benefits is also needed to cement consumer commitment and increase spending. Interestingly, although 59% of those surveyed by TNS agreed that 'organic foods are friendlier to the environment', only 48% declared themselves to be organic consumers. This suggests that belief in environmental benefit does not translate reliably into a readiness to purchase. When it comes to the arena of health, however – more personal and 'closer to home' – it would seem that virtually all those who recognised a health benefit (49%) did buy organic products. Market penetration and overall sales are likely to increase if the health arguments can be convincingly won and more widely disseminated.

This view of the pre-eminent importance of health motivations for organic consumers is reinforced by pan-European research conducted through the CONDOR project[6] and published in 2005. Researchers interviewed 8400 consumers in eight countries across Europe (Denmark, Finland, Germany, Greece, Italy, Spain, Sweden and the UK) and found that the reasons why consumers were buying were very similar. In all eight countries the belief that organic food is healthier was the strongest influence on purchase. There was also a recurring theme of positive moral values and peer esteem about eating organic. 'Organic consumers believe that by following their organic purchasing habits they are doing the right thing, and are seen to be doing so by friends, family and neighbours,' said the CONDOR project's summary brochure (see Figure 2.3). The bigger the coloured octagon on the grid, the greater the influence on consumer choice of that particular facet of organic food. The relative uniformity of the shapes shows that attitudes and their prevalence tend to be very similar from country to country – beliefs about organic food being healthier are the most important influence on choice in every country except Greece.

A taste to believe in

Consumer attitudes to taste were explored in a Market Tools online survey of 817 organic consumers in 2005.[5] Over 90% said 'enjoying a tastier diet' had been an important motivation for going organic. More than seven out of ten felt that organic fruit and vegetables tasted either 'quite a lot better' or 'much better' than non-organic, and there was a similar finding for organic meat. Organic fruit and vegetables had the highest taste rating, with 49% of respondents identifying them as the products with the most noticeable difference in taste from their non-organic counterparts. Organic meat gained 16% of the votes, eggs 15% and dairy products 6%.

The way in which many organic consumers swear by the taste of organic food reflects the fact that many organic products deliver on taste. The taste preference that consumers claim is, however, not consistently borne out in taste testing. Supermarkets report that faced by the objectivity of a blind taste test, organic enthusiasts do not

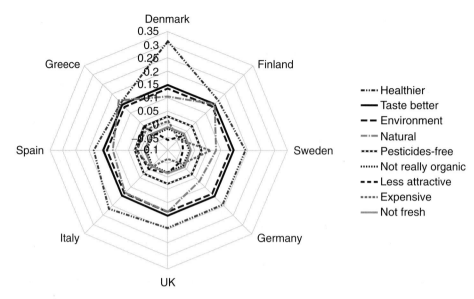

Figure 2.3   This diagram allows a visual comparison of consumer beliefs about organic food in eight European countries, based on data from the EU's CONDOR (Consumer Decision Making on Organic Products) project.

necessarily pick the organic option or even distinguish it from its non-organic counterpart. Qualitative research suggests that sometimes consumers ascribe positive taste attributes to organic foods where no appreciable difference necessarily exists – a consequence of a general 'feel-good factor' associated with eating organic. For example, focus groups conducted on behalf of the Soil Association in 2002[7] found that organic consumers enjoyed peace of mind from a food safety perspective ('You don't have to be scared of what you're eating for a change') and a sense of virtue and well-being ('You feel virtuous because you know you're not pumping your body full of antibiotics or whatever muck is in there'…'I feel I'm doing the right thing.'). Positive associations such as these may be influencing the emotional context in which tasting organic food is experienced.

Harriet Vogt, who conducted these focus groups, concluded that taste was a less direct and more subtle motivating factor than health for organic consumers, helping to 'reinforce belief'. Those buying organic food were doing so in the belief that it was something different, and this belief was then positively affirmed if there was a positive difference in taste. High-quality, flavoursome organic foods reassure those who have chosen organic products that they have made the right decision.

The pleasure principle

Whether a difference in taste is real or imagined, it is clear that the general feel-good factor associated with organic food translates into a genuine sense of enjoyment. Consumer polling in 2006[1] found that while 57% of those surveyed said they enjoyed organic and non-organic food equally, 38% found organic food more enjoyable and only 5% enjoyed non-organic food more. This contrast was even more pronounced among regular organic shoppers, with 54% saying they found eating organic more enjoyable, and only 2% saying this about eating non-organic products.

The 'negative' stimulus of food scares may have been the starting point for most organic consumers in the late 1980s and early 1990s, but the 'positive' sense of enjoyment captured by this poll has also played a significant part in developing the organic market in the past decade. A plethora of cookery programmes, magazines and newspaper sections has helped to revive interest in cooking and culinary tradition, using good-quality ingredients and discovering the provenance and 'story' behind food. Programmes such as Rick Stein's *Food Heroes* and *The Food Programme* on BBC Radio Four have thrust organic producers into the spotlight as beacons of traditional food culture, caring custodians of the landscape and animals and exemplars of artisan craftsmanship.

As the pace of life quickens, fuelled by bewildering technological development, instant communication and longer working hours, there is a wave of nostalgia in the midst of our convenience culture. Many are able and willing, however infrequently, to invest time and money in cooking from scratch with good-quality ingredients to make a family meal time that is a bastion of civility and communal enjoyment. Many organic producers, whose product often appeals to a yearning for 'how things used to be' and 'how food used to taste', are riding the wave and reaping the benefits.

The pleasures of good food have also come more to the fore in the marketing of organic food. The middle-market supermarkets, Sainsbury's and Tesco, have both identified 'foodies' motivated more by food quality and flavour than by ethical considerations as increasingly significant players in the organic market. Where previously the own-label organic products of these retailers communicated wholesomeness (Sainsbury's) or everyday value (Tesco), both have rebranded in a way that has given their products a more upmarket, aspirational, premium-quality feel and increased their foodie appeal.

The importance of foodie consumers can also be seen in the evolving branding of leading organic products such as Green & Black's chocolate and Rachel's Organic yogurt. In both these cases packaging has changed to emphasise premium quality first and organic status second. On each product the brand name has become more prominent than before and there is less emphasis on the word 'organic'. The 'O' word is not overplayed – it may carry the connotation of 'worthy but dull' for some foodie shoppers – but it is still there because it needs to be as a secondary reassurance of quality for some foodies and a primary badge of worth for those consumers most strongly motivated by health, the environment or animal welfare (see Figure 2.4).

Figure 2.4   The packaging for Rachel's Organic typifies how a number of leading organic brands in the UK have adopted less 'worthy' and more upmarket branding to appeal to foodies looking for premium food quality.

The extent of the food-buying public's aspiration for quality was underlined by a telephone omnibus survey conducted for the Soil Association by BRMB in 2005. When a representative sample of 1000 people was asked what was important to them when buying food for a meal to serve to family or friends, 95% said 'the taste and quality of the food'. Only 57% said low prices were important. Even among the least well off, quality and taste were considered important by 94% and low prices by only 65%.[5]

Food with a story

As the industrialisation of food production has alienated consumers and undermined their trust, the most successful organic brands and retailers have sought to build trust by telling the story of where their products come from. They have the distinct advantage that theirs is a story worth telling, rather than one that is better left untold by an anonymous, homogenous label.

The Soil Association's 2006 Market Tools/zOmnibus poll[1] suggested that all shoppers – and organic consumers in particular – were interested in where their food comes from and the story behind it. More than eight out of ten shoppers, and 91%

of regular organic shoppers, said they believed it was important for the country of origin to be stated on product packaging. Over half of all shoppers (52%) and four out of five regular organic shoppers (79%) agreed that 'I like the packaging to tell me about the farm a product comes from and/or the people who produced it'.

Given this appetite for information, it is unsurprising that organic shoppers are particularly fond of organic brands with a distinctive story or personality – 44% said they liked to buy 'distinctive organic brands such as Yeo Valley, Green & Black's, Clipper, Whole Earth and Grove Fresh', with only 19% expressing a contrary view. Only 13% felt that 'I have more trust in the supermarket's own organic products than in organic brand-name products', while 32% disagreed with this statement (see Figure 2.5).

Supermarkets and brands marketing organic products need to find a way to strike a balance between providing enough information about producers to create interest and establish trust in the product while paying heed to research that shows too much information on packs can confuse and demotivate. Point-of-sale display material and websites have an important role to play here, ensuring that consumers have access

Figure 2.5   Research suggests that organic consumers are particularly attracted to brands such as Clipper tea, with a distinctive story or personality.

to information but allowing the products themselves to carry straightforward labels. Following the lead of pioneering companies such as Graig Farm Organics (www. graigfarm.co.uk/traceability.htm), Sainsbury's has developed traceability codes and web pages (www.jsorganic.co.uk/trace.asp) that enable consumers to identify online where a product originated and read the story behind it.

Green & Black's is a good example of a successful company that has used the back of its labels to good effect, providing additional information about its cocoa producers in Belize without cluttering up the core branding. This less upfront approach to telling the product story seems to be vindicated by research for the Institute of Grocery Distribution (IGD), which found a very mixed response to the idea of personalising products to the extent of placing the producer's name or photograph on the pack. Of those surveyed, 14% had a very positive attitude to this kind of personalisation, and these respondents were more likely to be over 55, female, white and living in rural areas. Those with the most negative attitude were more likely to be in the 35–54 age bracket, wealthy, white and living in suburbia in the south of England. By making the personalisation of its products low key, a company such as Green & Black's can satisfy those hungry for information without putting off the sceptics.

Baby comes first

The birth of a child is often the trigger for buying organic food for the first time. By the end of 2004, sales of organic products accounted for under 2% of food and drink sales as a whole, but organic baby food accounted for 43% of the total baby food market.[5] The demand from parents for organic products is so strong that Sainsbury's has discontinued its own-label non-organic baby food range.

After the relatively carefree twenties, many first-time parents rethink their priorities when a baby comes along. The unique wonderment and sense of vulnerability evoked by a small child can awaken concern and a sense of responsibility about issues such as food safety and protecting health and the environment for future generations. The kind of views recorded by Harriet Vogt in her focus groups for the Soil Association[7] show how new parents think. 'There was the salmonella thing, wasn't there, when they were feeding chickens dead chickens,' said one. 'That's what scares you when you've got children. You're all of a sudden thinking: what on earth am I feeding them?' Another said: 'We are the generation that has been betrayed by the food industry … it's too late for us but perhaps we can protect our children.'

It is significant that the buying of organic food peaks among those aged 35–44 years, the age group most likely to have young children at home. Research in 2006[5] showed that 71% of people in this age group had bought organic food in the previous six months.

The success of organic baby food is a positive pointer to the future growth of the broader organic market. Many parents who have weaned their babies on organic baby food will not accept poor-quality food from nurseries, schools and children's ranges. First Learning was among the first companies to spot the market opportunities stemming from this, becoming the UK's first organically certified day nursery when it opened its doors to around 60 children in Shepperton, Middlesex, in 2004. Organix, another pioneering enterprise that produced the UK's first organic baby food in 1992, has now broadened its scope to create a successful children's range (Goodies – see www.goodies.uk.com), which has all organic ingredients, with no added sugar, colourings, preservatives or flavourings (see Figure 2.6). Center Parcs, one of the UK's leading holiday chains, introduced organic meals for children to most of its restaurants and cafes in 2005.

Many parents are actively involved in trying to put organic food on the menu in hundreds of schools around the country, outraged by the revelations about poor school meals in the Channel Four series *Jamie's School Dinners* and inspired by the healthier blueprint put forward by the Soil Association's Food for Life campaign two years earlier. Food for Life, launched in 2003, challenged the government to reintroduce strict nutritional standards in schools, restore proper food education and fund higher spending on ingredients for school meals. It also urged schools to adopt menus using ingredients that are 75% fresh and unprocessed, 50% locally sourced and 30% organic. The campaign and its inspiration, Nottinghamshire dinner lady Jeanette Orrey, were influential in prompting Jamie Oliver to make the series *Jamie's School Dinners*, persuading the government to change its policies and rousing the passions of parents to get involved in improving the quality of school meals locally.

Concerned parents are also creating a thriving market for organic hygiene, skin-care and clothing products aimed at babies and children. 'It's not surprising that when people become parents they become more worried about the environment,'

Figure 2.6 Organix, the company that pioneered organic baby food in the UK, has also been one of the first to develop healthier organic snack foods for older children with its successful Goodies range.

says Jill Barker, whose company Green Baby does a brisk trade in washable organic cotton nappies. 'They are suddenly aware of the effect the chemicals we use are having on the world our children are going to grow up in. It's not something that can be ignored any more.'

From fruit and vegetables to meat

If baby food is the organic starting point for the latest generation of children, then fresh fruit and vegetables are where it all begins for most adult consumers of organic products. Research in 2003[4] found that 55% of organic consumers tried fruit and vegetables before any other category. Eggs and dairy products were the second most likely categories to be tried first (see Figure 2.7).

The typical 'cycle of adoption' – the order in which consumers buy into different categories – is fruit and vegetables, then dairy and eggs, then 'ambient' grocery products, then meat. Shoppers can be encouraged to move through this cycle by cross-promoting some categories to the consumers of others. Go Organic, for example, promoted its pasta sauces via miniature recipe leaflets affixed to organic egg boxes. Tesco has promoted its organic grocery lines via leaflets in the packaging of fresh fruit and vegetables.

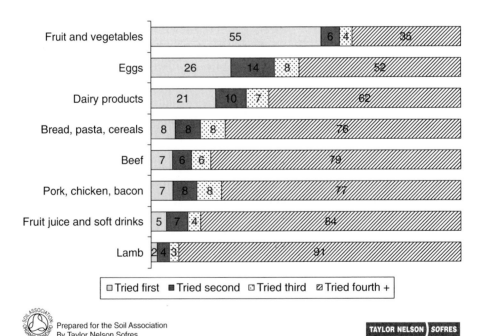

Prepared for the Soil Association
By Taylor Nelson Sofres

TAYLOR NELSON SOFRES

Figure 2.7   Fruits and vegetables are the most frequent entry point to organic shopping.

TNS has usefully examined how motivations differ for different kinds of foods.[4] These findings provide pointers as to which messages and issues to focus upon in promoting these products, although all the various motivating factors play a significant part in turning occasional shoppers into committed consumers.

- **Fruit and vegetables** – The main motivations were better taste (cited by 40%), avoidance of chemicals (37%), food safety generally (34%) and better quality (30%).
- **Eggs** – Animal welfare was uppermost (43%), followed by taste (37%), quality (32%), food safety (31%) and health (29%). Discrepancies between claimed purchasing of organic eggs and concrete spending data suggested consumer confusion, with rather a lot of people apparently buying free-range eggs because they thought organic birds were not free range or they equated one with the other.
- **Dairy** – Health predominated (46%), followed by food safety (34%), the environment (29%), food quality (28%) and avoiding GM ingredients (26%). Dairy consumers were more likely than the average organic consumer to have been influenced by a press article or shop promotion, and product quality was a key factor in driving consumers into heavy purchase. This was the only category where the environment was a significant motivating factor for more than a quarter of buyers.
- **Pork, chicken and bacon** – Health and animal welfare were the top motivators for light buyers. Taste was the most important factor for organic consumers overall (40%), followed by animal welfare (35%), food safety (33%), food quality (31%) and health (30%).
- **Packaged grocery products** – Press coverage and advertising were more likely to have been influential in persuading people to buy grocery products (e.g. bread, pasta and cereals), compared to other organic categories. The sequence of top-ranking motivating factors was taste (35%) followed by health (31%), quality (30%), additive avoidance (27%) and food safety (24%).
- **Fruit juice and soft drinks** – A particularly high proportion of purchasers cited concern about pollution, additives and the health of children as motivating factors. The sequence of motivations was taste uppermost (34%), followed by health (31%), quality (30%), additive avoidance (27%) and food safety (24%).
- **Beef and lamb** – Presumably because of BSE, food safety was a more powerful motivating factor with beef than for any other product. The top-ranking motivations for beef consumers were taste (38%), food safety (35%), food quality (31%), then health and animal welfare (both 29%). For lamb the sequence was taste first (38%), then animal welfare (31%), food safety and quality (both 30%) and health (29%).

The media and the message

Media coverage is more influential than any other channel of communication in driving the organic revolution. TNS reported in 2003[4] that 50% of people had first found out about organic food from editorial coverage in the print or broadcast media. Only 20% had found out from advertising, reflecting the fact that the organic 'industry' is still in its infancy and most of the companies involved do not have the resources for big advertising budgets. Word of mouth in the form of personal or professional recommendation were also important factors, with nearly one in six consumers saying that they first found out about organic products either from a family member or from a health professional (see Figure 2.8).

Print media coverage was found to have been particularly influential in relation to dairy and packaged grocery consumption; television coverage in relation to eggs; advertising for beef, packaged groceries and fruit juice; shop promotions for dairy products and fruit juice; and family influence for beef.

High-profile media coverage has been seen by the Soil Association to have a direct knock-on effect on sales of organic products. In Organic Week 2003, when publication of a hard-hitting report launched the Food for Life campaign and attracted widespread media attention, home-delivery box schemes around the country reported enlisting new customers.[9] In 2005, sales of organic dairy products rose by 50% in six months on the back of media coverage from scientific research publicised by the

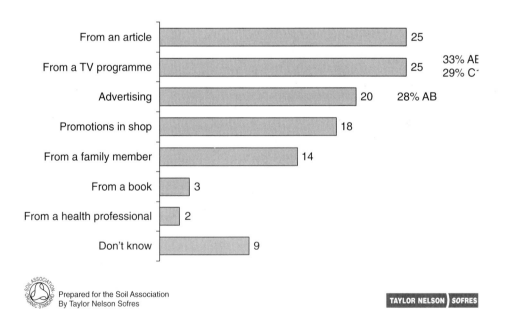

Prepared for the Soil Association
By Taylor Nelson Sofres

TAYLOR NELSON ❭ SOFRES

Figure 2.8   Answers to the question 'Where Did You First Find Out About Organic Products?'

Soil Association, Newcastle University and the Organic Milk Suppliers' Cooperative showing that organic milk contains significantly more vitamin E, beta carotene, omega 3 fatty acids and antioxidants (lutein and zeaxanthine).[10]

TNS research in 2003 provided plenty of information about the media preferences of organic consumers, who were less likely than the average viewer to watch BBC1, ITV1 and Channel Five and significantly more likely to watch BBC2 or Channel Four. The programmes found to attract significantly more interest among organic consumers than among viewers in general included drama, biography, music (especially classical), history, arts, news and current affairs. News (watched by 86%) and drama (85%) were the programme areas that reached most organic consumers.

The newspapers that had a significantly higher-than-average readership among organic consumers were all the daily former broadsheets, the *Daily Mail* and the *Mail on Sunday*, the *Sunday Express*, the *Observer* and the *Independent on Sunday*. Organic consumers were less likely to listen to Radio One and commercial radio stations than most people, but more likely to listen to the other four main BBC network stations, Classic FM and BBC local radio.

A matter of trust

While the organic market was largely kick-started by fear in the 1980s, producers seem to have earned the positive trust of a lot of committed consumers over time. In a 2006 survey,[1] 40% of the public at large said they had more trust in organic food than in non-organic products, with only 17% disagreeing. Among regular organic buyers, those believing organic food to be more trustworthy rose to 70%, with only 5% expressing a contrary view.

'You don't have to be scared of what you're eating for a change,' was one focus group participant's encapsulation of this sense of trust.[7] 'If it says "organic" I trust it because it can't go breaking some trade description thing,' said another. 'You're attracted to the word "organic",' said a third, 'because it does conjure up everything being natural, something you're looking to drive towards, a healthy way of looking at it. The word "organic" conjures up all those things in one word.' The CONDOR project[6] found that consumer trust was stronger, and negative associations with organic food weaker, in countries with more mature organic markets such as Denmark, the UK and Germany.

An important component of this trust is the independent verification, traceability and reassurance offered by organic standards and certification. According to focus group research in the UK, the certification mark on organic products is not something most consumers actively look for but there is a significant minority who do – usually the most committed shoppers. 'If it didn't have the mark I wouldn't trust it,' was a typical focus group articulation of this perspective.[7] 'I wouldn't believe it was organic. I'd think it was a rip-off.'

Among this group, the Soil Association's 'organic standard' symbol seems to be pre-eminent, with loyalists citing the more stringent standards it represents and a higher level of trust in the integrity of the mark.[7] Research among Sainsbury's, Waitrose and Tesco organic shoppers in 2005[11] found that this was likely to translate into some market advantage for Soil Association-certified producers. Of these shoppers, 84% said they would definitely or probably buy products bearing the Soil Association symbol compared to 74% for products displaying the European Union organic mark and 75% where a simple bullet-point motivational message about the benefits of organic farming replaced any certification mark.

A changing climate on the environment

Consumer attitudes to the environment are something of a conundrum for the organic movement. As we have seen earlier in this chapter, the benefits of organic farming to the environment and wildlife are among those most widely accepted by the public, and understanding of these benefits plays an important part in turning organic 'dabblers' into committed consumers. Forthright environmental messages may, however, actually be off-putting when it comes to attracting new consumers (see Figure 2.9).

Figure 2.9   The benefits of organic farming to the environment and wildlife are among those most widely accepted by the public.

Focus groups conducted by Corr Willbourn Research and Development in 2005 found that occasional organic buyers and the wider public had little personal engagement with the environmental agenda, and did not readily connect environmental issues with food and farming. If talking about the environment at all, some were most concerned about local community issues in the 'social environment' – litter, graffiti and feral youth. Others expressed concern about the effects of industrial development both at home (loss of greenbelt) and abroad (destruction of the rainforests). Even when they understood the causes and nature of threats to the environment, only very few claimed that they had changed their behaviour as a result. There was a sense that the problems were too big to grasp and were up to the government and business to fix.[12]

Hugh Willbourn, who conducted this research for the Soil Association, argues that simply communicating the environmental benefits of eating organic will not connect with many 'mainstream' consumers. Those in his discussion groups were very resistant to being 'preached at', and more likely to be turned off than turned on by what some felt were 'hippy' environmental messages. Climate change was the sole 'big picture' environmental issue that seemed to have greater salience and to be of fairly universal concern. No one connected climate change with the way food is produced, distributed and consumed, but this may change over time as the Soil Association and others publicise the connections. In its response to the Treasury review on the economics of climate change in December 2005, the Soil Association asserted that widespread organic farming would roughly halve the energy used in food production because it does not use synthetic nitrogen fertiliser. The Soil Association's submission also pointed out that nitrogen fertilisers are the largest source of carbon dioxide emissions in agriculture and their manufacture emits large amounts of nitrous oxide – another significant greenhouse gas (see www.soilassociation.org/sternreview).

The jadedness of mainstream consumer attitudes to the environment in Corr Willbourn's focus groups was unmistakable. Yet more recent quantitative research conducted on behalf of Seeds of Change[13] suggested that environmental motivations were more important than ever before in attracting new organic consumers. When asked which factors had 'strongly influenced' them to try organic food for the first time, 41% identified 'concerns about the environment' – making this the strongest factor of all.

So how can we explain this apparent contradiction? Perhaps the answer is that while environmental issues are so vast that they have the potential to disempower and demotivate, when it comes to buying organic food, increasing numbers of people are being attracted to and empowered by doing something positive that is personal, simple and practical. Perhaps the relentless reporting of climate change issues by the media is finally sowing the seeds of a more active environmental consciousness.

A mass of contradictions

Research shows organic consumers wrestling with their consciences as they face the ethical dilemmas of shopping for a better world.[1] Most succumb to the convenience of supermarket shopping, and many enjoy making savings from bargain prices and savouring their favourite foods from all over the world throughout the year. But many would also like to support their local farmers more, are concerned about whether farmers are getting a fair deal when they see cut-price offers, and want to cook and eat more seasonally.

Prices versus principles

Pricing is an area of particular soul searching. Price has emerged consistently in market research as the single most important factor deterring those who do not buy organic food. In 2006, Market Tools/zOmnibus found that 84% of people thought organic food was too expensive, and as many as 77% of those buying organic food at least once a month felt this way too. Yet 37% of the public as a whole and 63% of regular organic consumers agreed with the proposition that 'organic food tends to be more expensive but I think it is a price worth paying'. Nearly three-quarters of shoppers (73%) agreed that they 'like to see special low-price offers on organic food and drink'. Yet 30% of those surveyed (and 40% of regular organic shoppers) also agreed that 'low-price offers on organic products concern me because I worry about whether the farmer is getting a fair price', compared to only 20% disagreeing with this statement.[1]

Convenience versus conscience

Nine out of ten organic consumers do some shopping in supermarkets, compared to three out of ten in farmers' markets and two out of ten in farm shops. Only 43% identified supermarkets as their preferred outlet for buying organic products. When asked where they would prefer to buy if all outlets were equally convenient, 52% opted for smaller local suppliers such as greengrocers and butchers, farm shops, farmers' markets and home-delivery box schemes.[1] People clearly aspire to a leisurely shopping experience connecting them directly with farmers and to cooking from scratch with plenty of fresh ingredients, but time is at a premium in the busy dual-income household. The most successful smaller shops of the future may need to be like the Better Food Company in Bristol, offering the convenience of supermarket shopping (open until 7 PM each evening, all your household needs including convenience foods under one roof, bulk-buy savings) but with an outstanding ethical offer (a strong accent on organic and fair trade foods, unpackaged fresh produce, low food miles and no air freighted products, friendly service and free tastings involving local producers).

Local versus organic

Local food and seasonality are also issues where organic shoppers are wrestling with ethical dilemmas. There is something special about buying locally produced fruit and vegetables in season, yet imported out-of-season organic apples and asparagus fly off the supermarket shelves. When asked whether they would prefer to buy a locally grown non-organic product or an imported organic product, a clear majority of the public as a whole picked the local non-organic option. Supporting local producers and reducing environmentally costly 'food miles' were the reasons most frequently given for this preference. Among those buying organic food once a week or more support for local producers was still strong. But the proportion preferring the imported organic option was almost double that among the sample as a whole. Health was the biggest motivation for those expressing this preference (cited by 52%) but taste and the environment were also important factors (cited by 39% each). These findings underline the importance from an organic consumer perspective of continuing to develop organic production in the UK. The public will not need to choose between organic and local if they are given an increasing range of products that are both organic and local, optimising health and environmental benefits and salving the eco-consciences of thousands.

Beyond organic

There is much still to do for the organic movement if it is to educate a wider public, maintain trust and meet the ultimate expectations of its consumers. According to Hugh Willbourn, the sceptics still do not fully appreciate what organic means, while the enthusiasts now have aspirations that reach beyond what organic standards currently deliver. 'Close listening reveals that as a descriptor "organic" is not compelling and is minimally but not maximally understood,' says Willbourn. 'It does not evoke a clear sense of immediate, physical benefit and is not a salient differentiator.'

Among the most committed consumers Corr Willbourn found that 'organic' was far from being the only criterion for food purchasing. Food miles, packaging and ethical trade were all influential factors, with shoppers prepared to avoid buying some organic products if they perceived them to have been overpackaged or transported over excessive distances. Those in this group of consumers were found to have undergone something of a conscious and significant environmental lifestyle shift of which buying organic food was only a part. They were characterised by acute environmental awareness, tended to be interested and involved in local community issues and aspired to a different food culture. All shopped at farmers' markets, and some exclusively so.

In order to achieve the longer term aim of more organic consumers becoming 'lifestyle shifted' – in other words, thinking and acting very differently on environmental issues rather than 'only' buying a few organic products – Corr Willbourn urges a broadening of motivational messages. It advocates embracing localness and 'slow food' values and enabling more people to enjoy a wholly different experience in relation to food and shopping – the kind of emotionally engaging experience exemplified by the current 'buzz' around farmers' markets.

Building an organic future – the role of the Soil Association

The Soil Association has played a significant part in the growth of the UK organic market. An educational membership charity founded in 1946, it developed the world's first organic standards and works to highlight the impact of industrialised agriculture and promote organic food and farming.
   Its engagement with the organic consumer includes:

- A network of nearly 100 organic farms open to the public and to school visits across the UK. These farms attract over 750 000 visitors a year and enable the public to see organic farming for themselves. According to research in 2006,[1] 11% of people have now visited an organic farm, with over 99% finding it an enjoyable experience. Among those who have yet to visit a farm, 54% expressed an interest in doing so in the future.
- Publishing and publicising influential reports that have made the case for organic food and farming on environmental, health, animal welfare and social grounds, including *The Biodiversity Benefits of Organic Farming* (2000), which provides a comprehensive review of nine independent research studies comparing the variety and numbers of wildlife on organic and non-organic farms; *Organic Farming, Food Quality and Human Health* (2001), which collates and assesses the evidence for a wide range of health benefits associated with organic food; *Batteries Not Included* (2004), which highlights the animal welfare benefits of organic production and *Organic Works* (2006), which shows the increased employment potential associated with expanding organic farming. Also available are the annual *Organic Market Report* and the *Organic Directory* – a comprehensive where-to-buy guide.
- *Living Earth*, a three-times-a-year consumer and member magazine with a readership of 50 000.
- A wide-ranging main website, www.soilassociation.org, visited by around 70 000 people each month.
- A magazine-style, consumer-focused website, www.whyorganic.org, featuring product offers and competitions, basic information about the benefits of organic food and farming and the online version of the *Organic Directory* (see Figure 2.10).
- Support and advice for retailers, producers and food manufacturers in developing material to promote the benefits of organic food and farming to consumers. This has included painstaking research to assemble a growing list of claims

Figure 2.10 The Soil Association's 'why organic' website, launched in 2004, aims to help increase sales of organic food with a dual approach.

about organic food and farming which have been approved by the Committee on Advertising Practice for use in the promotion of products.[14]

- Organising national annual events such as Organic Fortnight (formerly Organic Week), the Organic Food Festival, the Organic Food Awards and the School Food Awards.
- A public information service that deals with enquiries from 25 000 people each year.
- Periodic joint promotional activities involving organic businesses, including the Organic Taste Experience sampling roadshow and money-off voucher booklets.
- Development of curriculum-linked schools materials including teachers' packs, *The Little Book of Organic Farming* and *Your Food, Your Choice* – a video aimed at primary school children.

Future trends

As Marketing Director of the Soil Association and Editor of its member magazine, the authors of this chapter have been in a good position to anticipate future market development through the regular research the organisation commissions and the pattern of enquiries received by its public information service and letters from members. To conclude the chapter, here are some pointers towards likely future consumer trends.

The challenge of local food

Food miles have become a 'sexy' issue in the media and middle England, a potent concept not least because it is an easy one to grasp amidst the complex science and language of the sustainability debate. Unfortunately it is also a concept that carries with it a danger of 'dumbing down' the environmental issues surrounding food production. Localisation is an important principle for the organic movement, but the more fashionable food miles become, the more attention will be focused on the blunt measurement of how far food has travelled, with comparatively little focus on the energy and natural resource use and environmental pollution associated with its production. To avoid being 'overtaken' by local food, the organic movement will need to increase UK organic production substantially and also to shift the public debate in a way that communicates the environmental credentials of all organic production, whether domestically or overseas. Banning air freight under organic standards – or introducing labels or web information that enable consumers who want to do so to avoid air freighted produce – is one way in which what is currently an environmental millstone for the organic movement could be turned into a milestone that distances it from the mainstream food industry.

Textiles, health and beauty and household cleaning

As recently as 2002, organic textiles were barely on the radar as far as consumers were concerned: 'Clothing was probably the lowest priority – so far from the source of production and having no immediate ill effect on the health of the wearer. Only one respondent was conscious of the scale of the cotton crop and its ecological effects and despite feeling that she ought to buy organic clothing, even she felt that organically grown cotton clothing tended to be unacceptably unfashionable.'[7] Now interest is booming and the range and quality of products have increased enormously. More and more consumers concerned about what they put into their bodies will in the future broaden that concern to encompass what they put on their skin, then what they wear (their 'second skin') and then the products they use in their homes ('third skin'). Research in 2006 found that the proportion of shoppers who had tried organic personal hygiene, beauty and cosmetic products did not exceed 10%, but over 50% said they would consider buying in the future.[1]

Convergence with fair trade

Consumers will increasingly expect fair trade products to be 'fair' in terms of their impact on the environment, while organic shoppers will want to know how 'organically' the farmers and growers producing organic food are being treated. When asked as part of a Market Tools/zOmnibus poll what they thought about the pay and conditions of those producing organic food, consumers clearly expressed an expectation that organic production and standards should take such issues on board. Four out of five shoppers – and 90% of regular organic shoppers – agreed that it was 'in keeping with organic values for farmers to be paid a fair price and for farm workers to be fairly treated and reasonably paid'. More than seven out of ten shoppers – and 83% of regular organic shoppers – said they would expect those producing a product labelled organic to enjoy decent pay and conditions. A similar percentage – 70% of the sample as a whole and 82% of regular organic shoppers – felt that this was a certification issue, agreeing that 'I believe, for a product to be certified organic, that the farmers and farm workers should be getting a fair deal'. Like consumers, producers are keen to see convergence between the organic and fair trade movements – not least because they are eager to enjoy the cost and efficiency benefits of combined inspection and certification. The Soil Association's fledgling ethical trade scheme has the potential to take off on the back of this consumer and producer interest (see Figure 2.11). This scheme provides consumers with independent verification that the farmers and farm workers making organic products have enjoyed decent pay and conditions, and is

Figure 2.11   BioBiz cereal, produced by Doves Farm Foods, was the first product to be certified under the Soil Association's 'ethical trade' programme.

an illustration of the potential for convergence between the organic and fair trade movements.

Differentiating primary produce

The Soil Association's Director, Patrick Holden, has enjoyed singular success by marketing the carrots from his farm in west Wales as 'single estate' produce in a branded bag sporting a painting of the farm by his daughter and the story of its 30-year organic heritage. These 'carrots with a story' have been outselling the standard organic carrots in Sainsbury's stores in Wales, and they show the potential for strong niche markets for differentiated primary produce in the future. Some growers will differentiate by locality and others by variety, capitalising on research findings associating specific benefits with specific types of crop. With livestock products there will be opportunities to differentiate by breed, promoting specific animal welfare benefits. As awareness and concern grow about the disposal at birth of unproductive male dairy calves, for example, a market could develop for milk specifically promoted as being from 'dual-purpose' breeds where male calves have a longer life through organic veal production.

Packaging and health

Some market-leading organic businesses and retailers have taken it upon themselves to start addressing the 'reduce, reuse, recycle' agenda in the way their products are packaged. Moves in this direction have tended to be driven as much by their own enlightened environmentalism and cost consciousness as by consumer pressure. In the future, packaging will become a more pressing issue for consumers, with an increasing bearing on shopping choices. The Soil Association information team reports that packaging has moved up the consumer agenda with media revelations about phthalates – making packaging a health issue (and therefore much more salient) as well as a matter of environmental concern.

References

1 Soil Association (2006) *Organic Market Report 2006*. Soil Association, Bristol.
2 Soil Association (1999) *Organic Food and Farming Report*. Soil Association, Bristol.
3 Department for Environment, Food and Rural Affairs (2002) *Action Plan to Develop Food and Farming in England*. DEFRA, London.
4 Welsh Development Agency/Soil Association/Organic Centre Wales/Agri-Food Partnership (2003) *Organic Food: Understanding the Consumer and Increasing Sales*. Taylor Nelson Sofres (TNS), London.
5 Soil Association (2005) *Organic Market Report 2005*. Soil Association, Bristol.
6 European Union (2005) *Consumer Decision Making on Organic Products*. EU Project Number QLK1-2002-02446. www.condor-organic.org

7   Harriet Vogt Brand Planning & Qualitative Research (2002) *Qualitative Research Analysis of Consumer Perceptions and Attitudes* (unpublished).

8   Smith, L. (2004) Children of the organic revolution. *The Independent*, 5 April.

9   Pearce, H. (2003) *Food for Life: Healthy, Local, Organic School Meals*. Soil Association/ Organix Brands, Dorset.

10 Nielsen, J.H. and Lund-Nielsen, T. (2005) *Healthier Organic Livestock Products: Antioxidants in Organically and Conventionally Produced Milk*. Danish Institute of Agricultural Sciences, Tjele.

11 Market Measures Ltd (2005) *Certification Logos Research* (unpublished). www.market-measures.co.uk

12 Willbourn, H. (2006) *Some Findings from Qualitative Research into Public Understanding of Organic Produce, Production and Related Issues for the Soil Association*. Soil Association Annual Conference, 7 January.

13 Ketchum Communications (2006) *Motivation to Try Organic Foods* (unpublished). Ketchum Communications Research Presentation. www.ketchumcomms.co.uk

14 Soil Association (2005) *What We Can Say – The Quality and Benefits of Organic Food*. Information sheet 24/11/2005 (version 4) available online at www.soilassociation.org/web/sa/saweb.nsf/b0062cf005bc02c180256a6b003d987f/ 7da7b6b517b1ba8280256fa50038c3ae!OpenDocument

Chapter 3

The Fairtrade Consumer

Harriet Lamb
Director, The Fairtrade Foundation

Additional research by Veronica Lasanowski

Introduction

All organisations have their legends: evocative stories about people that sum up
the spirit and purpose of the organisation and highlight what makes it special and
distinctive. In the UK Fairtrade movement, Angela Feaviour's story is becoming
legendary. In 2005, she asked the manager of her local pub in Leighton Buzzard
to stock Fairtrade tea and coffee. 'Sorry,' came the reply, 'it's not my decision. Try
the head office.' So Angela called the headquarters of the Slug and Lettuce chain.
Not only did she get agreement that her local pub would offer Fairtrade, the whole
chain would try it too. 'I found the right man, in the right place, at the right time
in the right frame of mind,' Angela recalls. Shortly afterwards the chain's parent
group SFI also adopted Fairtrade and by March 2006 were reporting double digit
increases in coffee sales.

It's not usually so easy. Normally, says Angela, getting companies to offer Fairtrade
products can be a slog. It is this hard work undertaken by thousands of campaigners, café by café, hairdresser by hairdresser, shop by shop, that has been critical to
the rapid expansion of the Fairtrade market, and is a unique characteristic of the
Fairtrade movement. Seeking first to build awareness about the inherent injustices in
the current system of international trade, the Fairtrade Foundation has actively fos-
tered what journalist John Vidal described as 'one of Britain's most active grassroots
social movements.'[1] By then encouraging consumers to look for the FAIRTRADE
Mark, the international certification mark guaranteeing a better deal for producers
in developing countries, the Foundation promotes Fairtrade as a viable alternative
to an unfair system of world trade. Campaigns have encouraged the public to use
their latent consumer power to tackle issues of global trade by buying Fairtrade
products, by telling family and friends about Fairtrade and by asking supermarket
and independent store managers to stock Fairtrade products. Thirdly, the Fairtrade
Foundation works in collaboration with companies which use traditional marketing
strategies to advertise Fairtrade products. While the importance of the companies'
own marketing cannot be underestimated, this chapter will focus more on those

marketing methods which are unique to Fairtrade and which seek to realise the core objective of bringing the organised consumer closer together with disadvantaged, organised producers. Together, these techniques have helped make Fairtrade fashionable, and it is this widening appeal that is examined first.

Cool to be kind: the evolving Fairtrade consumer

Since first appearing on products in 1994, Fairtrade has moved inexorably into the mainstream, and accordingly the Fairtrade consumer's profile has inevitably expanded and shifted. Today half the UK population – 52% in 2006 – recognise the cheery blue and green FAIRTRADE Mark (Figure 3.1). Sales of products carrying the FAIRTRADE Mark are rising in line with awareness, nearly quadrupling over the past five years in the UK, jumping from £50.5m (2001) to £195m (2005). The trend of 40–50% growth per annum is all the more noteworthy given that it is taking place in generally stagnant sectors such as coffee, tea or bananas. It is clear that today's savvy consumer is seeking out Fairtrade certified products. As a result, worldwide, some five million farmers, workers and their families across 58 developing countries are able to benefit from participating in, and shaping, Fairtrade.

The early adopters were, as could be predicted, predominantly older, better educated middle class women. While awareness remains strongest among women, AB classes and people between the ages of 35 and 54, that profile is widening fast. Repeated polling by the Fairtrade Foundation suggests that today's Fairtrade consumers are becoming younger and more diverse.

Indeed, with TNS Superpanel data showing that 40% of all households purchased a Fairtrade product in the past year, and with over 1500 products sold by 200 companies spanning the range from premium to everyday purchases, it is becoming impossible to pin down the Fairtrade shoppers. They are as diverse as the product offerings covering all ages and income brackets and with an even national spread. There are those who also loyally buy organic and free-range, shop in independent whole-food stores or more premium supermarkets and are concerned about the

Figure 3.1   The blue and green FAIRTRADE Mark is now recognised by 52% of adults in the UK.

provenance of food just as there are ASDA and Co-operative Group (Co-op) shoppers who believe the concept makes good sense. Any easy characterisation would therefore be as misleading as it would be quickly outdated.

It is also clear that Fairtrade is only one brand attribute: people also buy products because they like the quality, the taste, the style and if the price is right (which does not, of course, necessarily mean the cheapest), and if they can quickly find the products. Qualitative research confirms common sense: to succeed, Fairtrade products need to compete favourably against other products in terms, for example, of in-store visibility, quality and taste.

By 2006, recognition of the FAIRTRADE Mark was over 50% for every age group between 16 and 74. The highest recognition was among the 35–54 age group at 55%. However, some of the fastest growth was among the 16–24 age group, rising to 51% in April 2006, up 6 points on the year before.[2] That Fairtrade is becoming fashionable is clearly a positive sign for its future potential.

Across the social classes, significantly, people categorised as 'supervisory and clerical skilled labour' (C1) are becoming increasingly aware of the FAIRTRADE Mark with a 59% recognition level. Awareness remains highest – 66% – among people of 'managerial and professional labour' (AB).[2]

With 55% of Fairtrade buyers having first purchased a Fairtrade product in the past year, including 25% in the past six months, again these people who have more recently become aware of Fairtrade indicate the positive potential for Fairtrade sales in the future. Of those purchasing FAIRTRADE Mark products, 14% claim that their first Fairtrade purchase was in the first three months of 2006, with 5% in March (over Fairtrade Fortnight), and another 9% in January and February.[2]

In April 2006, 57% of women recognised the FAIRTRADE Mark in comparison with 48% of men, indicating that women do still have a slightly higher awareness of the mark.[2] This is in line with wider trends which show, for example, that younger women

---

Facts and figures: the Fairtrade consumer

- A 2006 TNS Omnimas poll shows that one in every two UK adults (52%) recognises the FAIRTRADE Mark
- According to a 2005 TNS research poll, Fairtrade consumers are becoming younger and more diverse
- People aged between 35 and 54 years are now most likely to recognise the FAIRTRADE Mark, with 55% of people correctly identifying the mark
- 55% of Fairtrade consumers first purchased Fairtrade products within the last year, including 28% within the past six months
- Supervisory and clerical skilled labourers are becoming increasingly aware of the FAIRTRADE Mark, with a 59% recognition level

are purchasing ethically sourced food in increasingly important numbers, indicating the growing influence they have on the total market. Their consumer clout may relate to the fact that women still tend to do more of the household shopping and in general have a greater interest in shopping. As a result, they are less driven by habit buying than men. A 2004 Mintel report about public attitudes towards ethical food suggests that women actually have a growing awareness about specific product labelling.[3]

While the Foundation's commissioned research polls indicate recognition and awareness of the FAIRTRADE Mark among adults (ages 16–74 years), the Foundation's work with younger people also indicates a very high level of enthusiasm for Fairtrade. The Young Co-operatives, for example, a nationwide network of co-operative working for students aged between 13 and 17 years, report the effectiveness of the 'pester power' of school-aged children as they encourage their family and teachers to buy Fairtrade. Young Co-operatives is a nationwide network that provides a practical introduction to Fairtrade and co-operative working. By giving younger people the chance to work in their co-operative, the program helps students to acquire valuable business skills and to make a difference in the lives of producers and their families. (For more information, see their website www.youngcooperatives.org.uk) Since 2002 the network has grown from just two registered co-operatives – the Chocolateers and Coco-Banana – to include more than 200 throughout England, Wales, Scotland and Northern Ireland.

As a result of programmes such as this, knowledge of Fairtrade among younger people is growing at an impressive rate.[2] This is set to rise further still as in 2006 the Fairtrade Foundation plans a national Fairtrade Schools Award Scheme to be implemented across the UK from primary to sixth form, thanks to a grant from the Department for International Development. The Young Co-operatives and the Fairtrade Schools Award Scheme appear in the List of Approved Projects in DFID's Development Awareness Fund. (For further information, see www.dfid.gov.uk)

The work of the Fairtrade Foundation in raising awareness of the FAIRTRADE Mark, and of the need for Fairtrade has undoubtedly contributed to wider changes in public expectations. A variety of independent polls confirm that consumers want to know more about the products they buy and they want to know more about procurement and standards for ethical production. For example, in 2004 Mintel research found that 79% of those questioned said ethical issues played a part when they did their grocery shopping. 'It is clear that an increasing number of consumers are thinking more about their relationship with the primary producers of the food they eat and about the conditions in which the food has been produced,' says Maria Elustondo, consumer goods analyst at Mintel.[4] This increase in concern also implies that media and PR campaigns have succeeded in raising awareness to a point that more consumers now feel confident in exercising their economic power when it comes to ethical issues'.[4]

In an annually published MORI poll on corporate responsibility, an overwhelming 94% of those surveyed think companies should report on their social and environmental impacts, with 59% strongly agreeing with the publication of company behaviour. Figures show that the public has high expectations for ethical standards in developing countries, with 90% of poll-takers believing companies have a responsibility to check that their international suppliers are behaving properly. More than 33% of people say they take a company's degree of social responsibility into account when deciding which products to buy, reflecting increased consumer responsiveness to ethical sourcing. People want assurances that the products they buy, and the companies behind them, are taking action to better socio-economic conditions, whether it is a guarantee against child labour or proof of fair wages for producers.[5]

The Fairtrade Foundation's own qualitative research confirms repeated findings that consumers are hungry for more information, and for clarification about complex ethical issues. Yet their desire to know more is balanced against the overload of information available. Consumers can feel all too like the proverbial rabbit in the headlights. In this position, they look to independent trusted sources of reassurance – and find helpful visual short cuts on packaging such as the FAIRTRADE Mark and bold images and stories.

Critically, independent polls commissioned by the Foundation found that four out of five consumers say that the independent guarantee of a fair deal for producers in developing countries is important to them, over half of them saying it is very important.[6]

We therefore have a position where the public are concerned about how their products are produced and they are looking for trusted, independent sources of reassurance. The key then is turning that evident concern into action.

Following the increased awareness of Africa during the Make Poverty History campaign, collaborative research between Marks & Spencer and the Fairtrade Foundation revealed that 83% of people believe the best way for businesses to reduce poverty in Africa is to improve long-term trade links with producers. (The TNS research was commissioned by Marks & Spencer using a telephone omnibus in July 2005.) Nearly 57% believe, moreover, that British businesses do not yet do enough to help African countries out of poverty. Most importantly, half of the British public say they would therefore buy more Fairtrade products to help communities in developing countries.[7]

Such research shows that these people are realising that their actions can have an immediate and positive impact on the lives of farmers in developing countries – that they, themselves, can help make poverty history. People are beginning to understand that 'by purchasing Fairtrade products they are making a real difference to growers and their communities,' comments Mike Barry, Head of Corporate Social Responsibility at Marks & Spencer.[8]

The Fairtrade movement undoubtedly has contributed to these attitudinal changes because it is a very positive and therefore empowering mechanism for consumers to realise change. According to the Ethical Consumerism Report 2005, people have become much more confident of their influence as consumers, with more than 50%

of people agreeing they can affect corporate responsibility.[9] The report found that in 2004, sales of ethical products and services as a whole increased by over 15% to total £25.8 billion.[9]

Getting into the habit

Overwhelmingly, people warm to the concept of Fairtrade and the reassurance of independent certification. Indeed, on the basis of focus group research conducted by Dragon Brand Agency, 'There are no barriers to the concept.' While many people have many questions about how Fairtrade works, practically no one ever disagrees with the fundamental proposition of a fair deal for poor producers. Both Dragon and repeated qualitative research have found that belief in the cause and the resulting emotional feel-good factor that consumers receive by purchasing a Fairtrade product is a key driver to grow the Fairtrade market.[10] Dragon aims to build better links between brands and the people with whom they are interested in connecting.

Focus group research by Research Works Limited found that the main motivation for buying Fairtrade was a willingness 'to do my bit.'[11] The research objectives of this report were to gain an understanding of what the existing FAIRTRADE Mark and identity convey to the general public, to understand the motivations and barriers to the purchasing of Fairtrade products and to find out where people expect to buy Fairtrade products. People felt good about buying Fairtrade, giving responses such as: 'It's doing something for somebody somewhere,' or 'If I saw my cream roses were Fairtrade, I'd feel twice as good buying them as an ordinary bunch', or it 'makes you feel good, if you feel you are doing your bit of good.'[12]

Qualitative research shows that while people feel that a positive attribute of Fairtrade was its uniqueness, it was an issue that should affect all shoppers, not only cause-minded ones. It should be 'everyday for everyone' not just 'premium'. They said it was 'powerful', 'meaningful' and 'important' because it was making tangible differences to farmers and workers in developing countries. The Foundation's own annual polling also shows that, critically, understanding of Fairtrade has risen in tandem with rising recognition. People do understand and support the principle proposition of Fairtrade.

As a result, purchases are rising strongly. Recent purchase rates are increasing year-on-year, with increasing loyalty as well as penetration key to future growth. The TNS Superpanel 2005 data show that 40% of households purchased Fairtrade products in the past year, recording an increase of over 21% on the previous year.[13] This was put down to three factors: increased penetration (more people buying Fairtrade), increased frequency (people buying Fairtrade products more often) and quantity (people buying more Fairtrade products when they shop).

The growing convenience of Fairtrade has been the convincing factor for those newer consumers, with continued innovation and product development likely to spur further momentum.[14]

Fairtrade is, however, very, very far from realising its potential. Barriers to consumer purchasing have been shifting over time, with quality perceptions, for example, becoming less of an issue (see below). Qualitative research reveals the key barrier now is simply that consumers are not yet aware of the scope and range of Fairtrade products – and would buy more if they knew about them. Focus group research by Research Works Limited found that for those already purchasing Fairtrade, the biggest barrier to buying more products was a lack of awareness of the range available. They said they tended not to 'branch out' into buying more and different Fairtrade items because they were not aware of the full range. Most people are aware of Fairtrade coffee, bananas and chocolate but products such as wine, honey or biscuits are less well known as yet (see Figure 3.2).

The majority said the influence of habit was strong: 'You tend to get stuck in your ways'. Many people admit that they buy according to long-established taste and brand preferences. Interestingly, this research found that men were particularly wedded to 'their brands' while women were more likely to experiment.[11]

This qualitative research was confirmed by TNS research in 2006 showing that the British public tends to stick to old habits when they shop, often purchasing familiar and well-known brands.[2] Only 3% of those who rarely or never buy Fairtrade cite quality as the key obstacle to purchasing products with the FAIRTRADE Mark. It is availability and in-store visibility of products that prove the more problematic issues

Figure 3.2 This photo highlights the growing range and availability of Fairtrade products. Fairtrade products now include items from coffee and cocoa to footballs and cotton socks.

for consumers. The most common reason given for not purchasing more products carrying the FAIRTRADE Mark is that they are simply not visible enough when out shopping. Over one third of those asked who knew about Fairtrade cited this as the biggest barrier to increasing Fairtrade purchase, while one in five (20%) admitted that they were simply not yet 'in the habit' of buying them. This result again confirms qualitative research showing that people need Fairtrade products to be promoted more effectively in-store if they are to find and so purchase them.

So we are it seems, creatures of habit, spending mere seconds in the coffee aisle before reaching for old favourites. TNS shopper data also shows that people may be buying one Fairtrade product – bananas, for example – but no products in other categories, such as Fairtrade coffee or tea.[13] Or people are buying Fairtrade occasionally but not regularly. As a result, in 2006 the national promotional activity, Fairtrade Fortnight, focused on a call to action, to 'Make Fairtrade Your Habit.' By explaining how a small change in shopping can have a major impact on the lives of producers, the explicit objective of Fortnight 2006 was to build on the progress made in 2005 in highlighting the growing range of UK Fairtrade products – 1500 products from 200 companies. Encouraging retailers to draw consumer attention to Fairtrade products on the shelves was, of course, critical to the campaign's success.

This focus on making it easier for people to take up new habits fits with recent findings from the Sustainable Consumption Roundtable (a group of experts in consumer policy, retailing and sustainability who advised the Government on consumer choices and environmental limitations from September 2004 to March 2006), which are summarised in the report's title: 'I will if you will: towards sustainable consumption.'[15] The report notes that the 'key to achieving one-planet living lies in making sustainable habits and choices easier for us to take up ... We need to feel confident that we are acting in step with others – neighbours and colleagues, friends and family – not alone and against the grain.' (The full report can be downloaded from www.sd-commission.org.uk or www.ncc.org.uk.) This is precisely the philosophy that the Fairtrade Foundation and its licensees have also been seeking to develop – positioning Fairtrade as the special part of the everyday shopping, an easy action everyone can take that will contribute to global change.

Because Fairtrade is a broad concept and practice focused on tackling poverty and promoting sustainable development, it fits with different brands and products in different ways; indeed, across the 200 licensees there are companies targeting all segments of the market. Both Marks and Spencer and the Co-op, two very different retailers, have switched all their own-label coffee to Fairtrade, establishing it firmly as a concept that spans the range of shoppers – from those seeking instant everyday coffee to those seeking organic roast and ground. In May 2006, Tesco reported on their website that every week 25 000 Tesco customers were buying Fairtrade products for the first time while the number of Clubcard customers buying three or more Fairtrade products each month nearly trebled over the past two years – evidence of all that has been achieved to date but also more interestingly of future possibilities.

Focus group research by Dragon found that people often compared Fairtrade with organic but did not directly associate with it. In other words, they were not confused themselves between the two issues, although packaging showing the two certification marks did cause some confusion.[10] In fact, across the USA and Europe, Fairtrade products are frequently also organic. In the UK, however, the Fairtrade Foundation has worked hard to ensure that the two schemes are clearly distinguished so as to avoid confusion and overly high expectations, and so as to ensure that smallholders who cannot for different reasons convert to organic can still benefit from Fairtrade.

When it comes to price, there are clearly two strong and paradoxical trends affecting the UK public. On the one hand, as we become wealthier and wealthier as a nation, we are spending less and less of our income on food – down from one-third in the 1950s to one-fifth now. Indeed, today's average household spends more on leisure activities than it does food and drink; the Office of National Statistics reports that while 17% of total expenditure went on food and non-alcoholic drink, 18% went to leisure. (These figures are based on statistics for 2003–2004 published by National Statistics Online. Leisure goods and services were the largest item of expenditure for UK households (£76 per week), while households spent £65 per week on food and non-alcoholic drink. For further information, see National Statistics Online: www. statistics.gov.uk.) To give just one example, the price of loose bananas in the UK have dropped by 42% between 2002 and 2006 from £1.10 to 64p per kg.

On the other hand, the public are ready to pay premiums for products they value highly. This includes a willingness to pay a premium for Fairtrade. In particular, when they hear of the consequences of this drive for ever lower prices, the public are uncomfortable with the results. No one likes to hear of thousands of banana smallholders and workers losing their livelihoods so we can save pennies. Hence shoppers are also on the whole ready, and indeed willing, to pay more for products with the FAIRTRADE Mark – and expect to, as it is consistent with the fundamental proposition that the producers are paid fairly. The challenge for Fairtrade is the wide variations – and often catastrophic falls – in the prices of mainstream products, for example, when supermarkets slashed 25% off the prices of ordinary bananas from 85p per kg to 64p per kg in March 2006, so greatly widening the gap between these and the stable and sustainable prices of Fairtrade bananas.

On ethical issues (as opposed to actual products), the public as a whole is wary of supermarkets and major corporations and they seek the reassurance that the premium charged for Fairtrade is reasonable. There are some who expect the simpler, more charitable proposition that '10p extra on this pack goes to the farmer'. However, when it is explained that Fairtrade is a long-term scheme to make trade fairer and to develop more equal relationships that move away from charity and are therefore sustainable, most people in fact prefer the concept of enabling producers to stand on their own two feet – of helping them to help themselves.

In summary, over the past decade we have witnessed fundamental shifts in public expectations of company behaviour and, consequently, in shopping habits. These shifts have resulted from many, many factors, but the work of the Fairtrade

Foundation and its key partners, from non-governmental organisations (NGOs) to licensees, has undoubtedly played a contributory part.

The Fairtrade Foundation

In 2005, the Fairtrade Foundation was voted Britain's Most Innovative Charity in the Third Sector Magazine awards for its 'trail-blazing role'. One reader noted that not only has it changed the face of UK food retailing, but it has also broken down significant barriers between charity and commerce. 'It has cracked this combination of charitable activity and trading for sustainability in a way that others have talked about but few have found a way of doing,' says Cliff Prior, chief executive of UK charity Rethink. [16]

A uniquely eclectic mix of innovative promotional strategies has contributed to the rise in Fairtrade awareness and sales. There can be few products in this country which are promoted by such diverse means. Across the country people have organised floats at Notting Hill Carnival or banners waving at Make Poverty History marches, to church coffee mornings organised by the National Federation of Women's Institutes (NFWI), African farmers speaking in town hall after town hall and school kids running tuck shops in their break times (see Figure 3.3). Alongside

Figure 3.3   The high level of grassroots activism represents a unique aspect of Fairtrade. (Photo courtesy of SCIAF.)

these grassroots methods, Fairtrade also benefits from completely conventional, highly expensive textbook marketing techniques of TV and print advertising, point of sale and in-store promotions by major listed companies, in addition to a raft of original strategies falling between the two extremes and at their most interesting, drawing on the best of both.

The success of these strategies can be judged by the rise of Fairtrade from fringe to mainstream in a relatively short space of time – the first three products carrying the FAIRTRADE Mark being launched in 1994 – with extremely limited resources. Yet by 2005, sales of Fairtrade products were doubling every two years, beginning to notch up serious market shares (see Figure 3.4). For example, some 20% of all roast and ground coffee in the UK carried the FAIRTRADE Mark while in Switzerland 47% of bananas, 28% of roses and 15% of pineapples were Fairtrade. By 2006, UK sales of products with the FAIRTRADE Mark were running at an estimated annual retail value of £200m. These sales delivered benefits to over 5 million farmers, workers and their families in 300 producer groups across 58 countries in the developing world.

In September 2005, the Foundation's achievements in building consumer awareness of the FAIRTRADE Mark were recognised by the Superbrands Panel of brand experts. In a YouGov public poll for the Superbrands Special Recognition Prize in the Media and Services Category, the FAIRTRADE Mark was named the winner, beating such well-known names as AOL, BT and The Times. Chairman of the Superbrands Council, Stephen Cheliotis, says that the 'FAIRTRADE Mark has gained considerable media and consumer awareness in recent years, owing to the persistent efforts

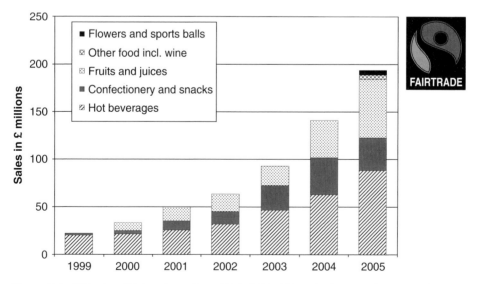

Figure 3.4   UK sales of Fairtrade products 1999–2005. 2005: 38% increase from 2004. Sales of Fairtrade products in the UK have been rising at an impressive rate during recent years.

of its founders and supporters. The steady increase of products and brands bearing the FAIRTRADE Mark illustrates the success of the Foundation.'

He continued: 'The Foundation is not only innovative in its aspirations, but also through its promotional work, including the highly successful annual Fairtrade Fortnight, which provides a great platform for essential promotional work. This commitment to genuine fair trade values alongside unique campaigns has earned Fairtrade the highly acclaimed accolade of Superbrand 2005.'[17]

A strategy of collaboration

The Foundation's core strategy of working with, and through its non-governmental owners, means that since day one it has been able to reach out to millions of potentially interested people – and through them to even more. For example, in the early days, Christian Aid called on its supporters to collect till receipts and send them to supermarket managers so as to encourage them to stock more Fairtrade products, providing a classic example of effective collaborative campaigning.

The Fairtrade Foundation was set up in 1992 by CAFOD (Catholic Agency for Overseas Development), Christian Aid, Oxfam, New Consumer, Traidcraft Exchange and the World Development Movement (WDM), with the NFWI (National Federation of Women's Institutes) – Britain's largest women's organisation, with a membership of around 250 000 – joining shortly thereafter. From the outset, these organisations gave the Foundation its moral authority, its credibility and its base in civil society, and they remain central to the trust placed in the FAIRTRADE Mark.

The Foundation later opened up its ownership, encouraging organisations with a strong development or consumer focus or history of supporting Fairtrade to join. As a result, key partners such as People & Planet, Banana Link, Nicaragua Solidarity Campaign, United Reformed Church, Scottish Catholic International Aid Fund and Methodist Relief and Development Fund all became 'charity shareholders' in 2003, with the Shared Interest Foundation joining one year later. Together, these organisations, many with significant mass memberships or nationwide group structures – Oxfam, for example, has upwards of 600 000 supporters – have been critical to raising awareness about Fairtrade, as they first informed their own members, who in turn often campaign locally for Fairtrade. It is only more recently that the organisation has supplemented this collaboration by developing its own supporter base.

Case Study One: NWFI

In 2001, a resource pack was developed in collaboration with the NFWI. Central to the pack is a video (*Fields of Gold?*) following two women members on a trip to the Windward Islands where they meet banana farmer Nioka Abbot and fellow farmer Betty Williams from St Vincent and even bravely try their hand at digging the farms. Through their eyes, we learn about the methods by which banana farm-ers are organising into Fairtrade. Along with background information materials, the video was sent to local NFWI groups, reaching thousands of women across the nation and in so doing, providing them with a ready-made kit about Fairtrade. The NFWI later developed their own campaign linking Buy Local – with their traditional support for local farmers – with Buy Fairtrade.

Case Study Two: Oxfam

Another major campaign was Oxfam's 'Coffee Rescue Plan' campaign of 2002. Oxfam focused on problems in mainstream trade, highlighting the crisis in low coffee prices and the damaging policies and practices of major multinational companies in particular. A weighty report detailing the crisis facing world coffee growers, 'Mugged: Poverty in Your Coffee Cup,' explains how falling coffee prices affect the livelihoods of 25 million coffee producers around the world. After explaining how the price of coffee had fallen by almost 50% between 2000 and 2002 to a 30-year low, the report proposed a 'Coffee Rescue Plan' to restabilise the market by encouraging the purchase of Fairtrade coffee. Backed by this report, the campaign called on the public to use their purchasing power as a positive tool for change, and challenged the major coffee roasters to clean up their act, including with an offer of Fairtrade products. At this time, in a ground-breaking move, the Co-op converted all its own-label coffee to Fairtrade, published its own report, and partnered with Oxfam to urge the major roasters to follow their lead by making the switch.[18]

This is collaboration at its best: Oxfam as an independent charity and household name with expertise and credibility campaigning for major changes in the whole sector, indirectly raising awareness of the need for Fairtrade and actively encouraging the public to buy Fairtrade, with the Co-op and leading brands such as Cafédirect showing both what companies can and should achieve and offering the public an easy means to play their part in creating change here and now while helping shift the whole coffee industry.

The campaign undoubtedly played a role in helping boost sales of Fairtrade coffee. At present, the British public consume more than 4.3 million Fairtrade hot drinks each day. Cafédirect, the UK's largest seller of Fairtrade hot beverages, is now the sixth largest coffee brand in the UK, with retails sales having reached £24.1m in 2005, an increase of 12.5% from the previous year (£21.5m).[19]

For members of NGOs, churches and trades unions such as Unison, Fairtrade has proved extremely popular, providing lay people an easier entry point to the complexities of international trade campaigning. It is not easy for campaigners to engage busy shoppers in discussions about the intricacies of world trade rules. It is extremely easy, however, to offer sample pieces of luscious Fairtrade chocolate. Many groups report that people often become interested in campaigns for wider political change after learning more about the stories behind the products filling their shopping baskets. This ease of effort helps to explain the popularity of Fairtrade with the local campaigners who have kept Fairtrade on the national radar screen of major development NGOs.

Fairtrade towns

'Fairtrade's work is both immensely practical and truly geo-political,' explains Tony Hawkhead, Chief Executive of environmental charity Groundwork, a supporter of the Fairtrade Foundation as Most Innovative Charity. Chief Executive of Victim Support, Dame Helen Reeves, pointed to the importance of the Foundation's increased profile in a competitive market: 'This has brought it to a much wider range of customers, contributing to public education as well as increasing the benefits for the producers.'[16]

While the Foundation's strategy has been first to reach people already organised into groups, it has since developed its own organised base. By 2006, the Foundation had over 63 000 supporters receiving its thrice-yearly published newsletters; some 5800 on an activist database receiving more frequent emails and updates; and over 10 000 people on a multiplier mailing for events organisation containing Fairtrade news and product updates.

The Foundation has also supported the burgeoning Fairtrade towns movement. By mid 2006, 200 towns, villages and major cities across the nation – from the appropriately named Fair Isle and Edinburgh City to Welsh and Cornish villages – were declared to have reached the Fairtrade goals, with at least another 200 in the pipeline. London is working towards becoming a Fairtrade city, borough by borough, and the Foundation is hoping London could be the inaugural city to host the world's first Olympic Games in 2012 where Fairtrade food and drink is available at official catering outlets.

The campaign was launched in 2001 when Garstang declared itself the world's first 'Fairtrade town' and called on other towns to follow suit. The Fairtrade Foundation set five goals that have to be met before a community can be recognised as a Fairtrade town:

- The local council must pass a resolution supporting Fairtrade, and serve Fairtrade coffee and tea at its meetings and in offices and canteens
- A range of Fairtrade products must be readily available in the area's shops and served in local cafés and catering establishments (targets are set in relation to population)
- Fairtrade products must be used by a number of local workplaces (estate agents, hairdressers, etc.) and community organisations (churches, schools, etc.)
- The council must attract popular support for the campaign
- A local Fairtrade steering group must be convened to ensure continued commitment to Fairtrade town status

Bruce Crowther, a veteran campaigner from Garstang, is now employed by the Foundation to coordinate Fairtrade towns. 'Most people are decent and if they understand what's happening in the world they will support Fairtrade,' notes Crowther. 'But there is such a swirl of information in the modern world and the hard part is getting people to take the time to listen and to wake up to it. That's where Fairtrade Town comes in. It brings the community together. It gets people asking questions. It gets into the newspapers and local radio and into schools and churches.'

While the long-term aim is to eradicate global poverty, the concept of the Fairtrade town works precisely because it allows people to achieve bite-sized goals rooted in the local community. As a consequence, it is a positive campaigning experience that energises people to take on the next step and the next. A street survey in Garstang a year after it became a Fairtrade town showed that more than 70% of people recognised the FAIRTRADE Mark, a figure significantly above the national average, which was then only 20%.

Thanks in large part to a campaign by the student group People & Planet, there are now 29 Fairtrade universities. Meanwhile, church-based organisations such as Tearfund and Traidcraft have successfully helped to establish 2000 Fairtrade churches. Synagogues, mosques and temples are likewise beginning to get involved in the Fairtrade movement, and as the Foundation works with education-based groups, a network of Fairtrade schools is flourishing.

Clearly, the Foundation is succeeding precisely by maximising the effectiveness of this grassroots strategy. Word of mouth and community networks have added value, local flavour and personality to the range of networking strategies and promotional offers developed by the licensees and retailers themselves.

The most concrete indication of the public's growing familiarity with Fairtrade is the Foundation's annual poll on awareness of the FAIRTRADE Mark. This shows that of those half who recognise the mark, an impressive one-fifth first heard about it through word of mouth.

The broadening appeal of the Fairtrade story has also contributed to rising year-on-year media coverage. Fairtrade has human interest, business and trade interest, consumer appeal, political and economic reach, and an international development agenda, but it also commands a strong local presence, simultaneously pushing a wide variety of buttons for media outlets. The local activities of the Fairtrade towns and universities and churches give the story – which is international in intent – a local angle.

Media stories thus bring Fairtrade alive for the majority of the public, with 2006 TNS data showing that 25% of those surveyed first heard about Fairtrade from the media. During Fairtrade Fortnight 2006 alone, there were well over 2600 mentions of the Fairtrade Foundation and the FAIRTRADE Mark in the press, enabling more people than ever to hear the story. The Foundation and the licensee companies have also worked hard to ensure journalists have the information they need and above all, the chance to visit producers in developing countries whenever possible so that they can see first hand the impact of Fairtrade for the farmers and workers themselves.

The Foundation welcomes press queries from a wide variety of sources, from popular consumer press such as women's magazines or lifestyle TV programming, to press specialising in trade issues. Such outlets may be less likely to cover international development issues normally, but their readers and viewers are interested in the goods they buy and the stories of the people behind them.

For example, ahead of Fairtrade Fortnight 2006, the Fairtrade Foundation facilitated a trip to Ghana in collaboration with the Day Chocolate Company and AgroFair after being approached by morning television programme GMTV. This trip resulted in a 30-minute Special Report on Fairtrade, aired over two days at the beginning of Fairtrade Fortnight. Thanks to the popular style of GMTV, the programme covered Fairtrade in an accessible and very comprehensive way, with Lorraine Kelly chatting on the sofa with Comfort Kwaasibea, a cocoa farmer from Kuapa Kokoo one day, and celebrity Gail Porter, the next.

The FAIRTRADE Mark

While the Fairtrade Foundation's overall vision is to promote a general awareness of Fairtrade and to encourage the purchase of Fairtrade products – thereby boosting opportunities for disadvantaged farmers and workers in developing countries – a more specific objective is to focus on the critical role of the FAIRTRADE Mark as an independent consumer label. Every year the Foundation tracks not only rising awareness, but also checks if people correctly associate and so understand the

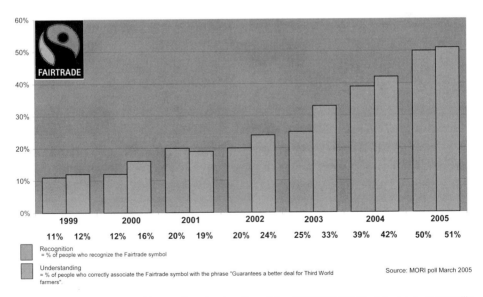

Figure 3.5 Public recognition and understanding of the FAIRTRADE Mark has grown steadily and reached 50% in 2005.

FAIRTRADE Mark. Each year understanding has risen in tandem with recognition (see Figure 3.5). In the very early years, the public as you expect overwhelmingly associated the FAIRTRADE Mark more with the Office of Fair Trading, but by 2006, 51% of people correctly associated it with a better deal for producers in developing countries.

However, the Foundation's own research in the early 2000s also showed that beyond the general grasp of the purpose, the public did not always understand precisely what the FAIRTRADE Mark stood for, while core supporters were often hungry for more information about the certification system and its guarantees. In response, the Foundation therefore sought to develop more detailed information for those willing to engage further, and in 2005 communications work focused explicitly on the guarantees behind the FAIRTRADE Mark. In 2006, the Foundation has also developed strategies to underscore its own identity, as the independent charity behind the FAIRTRADE Mark and the leading authority on Fairtrade in the UK.

The role of the FAIRTRADE Mark is especially important in Fairtrade, as compared to organics, for example, because there is no government legislation defining Fairtrade. The Foundation therefore places considerable stress on the importance of the FAIRTRADE Mark and its visibility on the front of product packaging. In this way, the consumer can be readily reassured that a product really has met independently set and monitored standards of Fairtrade and is not simply a company's own claim. The risk of companies making their own multiple claims without independent certification is that in the end the public become first confused and then cynical, to the detriment of all. Hence public trust in the single FAIRTRADE Mark is vital to maintaining consumer support. With a rising number of different labels covering different issues, the single FAIRTRADE Mark helps at least keep clarity in relation to trade with developing country farmers.

Working with licensees and retailers

As a certification body and charity, the Foundation remains vigorously independent of the companies. The ultimate aim is clearly to encourage the public to buy Fairtrade products so the producers can benefit and therefore all of the Foundation's work also leads people to the products and encourages regular purchasing. This ranges from profiling products in materials, for example, or directing them to associated web links to developing strategies in collaboration with the companies that are officially licensed to use the FAIRTRADE Mark on one or more products. Numbering more than 200 by mid 2006, these companies include both those offering only a few Fairtrade products and dedicated pioneer brands committed to a 100% approach to Fairtrade such as Equal Exchange, Traidcraft, Cafédirect and Divine chocolate.

These licensees employ a full range of marketing strategies and approaches, each of which targets different sections of society depending on their products. Some prefer to highlight the top quality of a Fairtrade product, others run special promotions.

With the fastest growing supermarket sales of Fairtrade fresh produce, ASDA, for example, ran special promotions on Fairtrade oranges, grapes and mangoes during March 2006. Waitrose, meanwhile, switched all their Caribbean loose bananas to Fairtrade and promoted the switch with a display showing a prominent photo and quote from a Windward Islands banana farmer, stressing both personal touch and provenance.

Other companies focus very much on Fairtrade as their unique selling point. Dubble, for example, hosts a website that highlights the difference Fairtrade makes to producers: 'It's a joy, and cocoa farmers feel really proud. They never thought they would own a third of a company in London,' enthuses Mr Ohemeng, Managing Director of Kuapa Kokoo Limited in Ghana, who supply Dubble with Fairtrade cocoa. Interestingly, the Dubble website first set out to educate kids, developing chocolate bars as the vehicle by which to do this. A direct collaboration between Kuapa Kokoo and two NGOs – Christian Aid and Comic Relief – the company works hard to help children appreciate the realities of life for the world's cocoa producers: 'Cocoa farmers need money to pay for essential things like medicines, school fees to send children to school and farm tools like Wellington boots to protect them from the snakes and scorpions that live among the cocoa trees.' (For further information, see the Dubble website: www.dubble.co.uk.)

An example of a company using the full range of strategies is the Co-op. Brad Hill, Consumer Policy Manager, points out that as a membership organisation, the Co-op adds an extra dimension to its commercial work by investing significantly in support towards Fairtrade towns and schools. 'What we needed to do,' says Brad, 'was to build awareness and to build the availability and then the visibility of products and thereby build sales. It was about developing products and developing communications. The two go hand-in-hand. The advertising is as important as product development. There's no point putting products on the shelves if you don't tell people. Equally there's no point in shouting about Fairtrade and not making products available. It's been a very managed process.'

When the Co-op converted all its block chocolate and then coffee own-label ranges to Fairtrade, it was backed by investing heavily in both print and TV advertising as well as in high profile media work challenging other companies to follow suit. After its chocolate initiative, for example, the Co-op posed a series of questions about the conditions of cocoa growers, chocolate prices and the FAIRTRADE Mark on its website. The results were telling; 98% thought all major chocolate manufacturers should make at least one FAIRTRADE Mark product; 87% considered or cared about conditions of cocoa growers when purchasing their favourite chocolate bar, and 94% would be prepared to see a slight increase in price if the chocolate was Fairtrade. (For further information, visit www.co-op.co.uk.)

'We tried to stir things up and champion Fairtrade as we went along. It was a leap of faith but we maintained the commitment and continued to build the Fairtrade market,' says Hill. 'The great thing for Fairtrade is that we've proved there's a marketplace out there that is mass market. It's not just for the affluent. The catchment area is almost completely irrelevant. We do a lot of trade fairs and for every student who says "I can't afford it, I live on beans," there is another who says, "I'll find the money because I support it and I believe in it."' Data from a 2006 TNS Worldpanel report shows that while the Co-op has 4.8% of the total food-market, it has captured 19.1% of the Fairtrade market, clearly demonstrating the success of their support for the Fairtrade movement.[20]

Fairtrade Fortnight

The Foundation's broad strategy of collaboration with all its key stakeholders has also been at the heart of Fairtrade's major push, the annual Fairtrade Fortnight, a two-week period of Fairtrade campaigning and promotion which has taken place in early March for the last 10 years.

The Foundation estimates that in 2006 nearly 10 000 events were organised by supporters across the country – in workplaces, clubs, universities, cafés and restaurants, shops and supermarkets, churches and other venues. The Foundation's staff devise the concepts and materials to be used by local supporters, while producers tour the country in order to speak at numerous meetings in support of Fairtrade. Licensees and retailers also promote Fairtrade in a myriad of other ways, from point of sale markers to car park and window posters, special displays and promotional offers. In 2006, Sainsbury's in-store promotions included triple Nectar card points on FAIRTRADE Mark products while Tesco ran in-store taste tests and targeted money-off vouchers to Club Card buyers of Fairtrade.

It is the combination of activities that is most effective. Someone could hear about Fairtrade on the morning TV, have the opportunity to talk to a Fairtrade supporter on a stall in the high street, and then easily find products in the growing number of whole food and Fairtrade shops and cafes. Many companies and retailers seek to launch major new initiatives in Fairtrade during the Fortnight since this is the period in which they have the best chance of catching the attention of the public and press. In 2006, Marks & Spencer put socks containing Fairtrade certified cotton on its shelves and announced that it was switching all of its tea and coffee to Fairtrade. Likewise, Virgin Trains announced that it would be switching all of its tea, coffee, hot chocolate, sugar and even chocolate sprinkles, to Fairtrade on board.

Sales of products carrying the FAIRTRADE Mark often treble during the Fortnight providing vital impetus to the continuing growth. The key is then to maintain that momentum and the Foundation is seeking to develop other key activity points through the year. In 2006, for example, an autumn campaign will focus on encouraging people to switch to Fairtrade in their workplaces.

Celebrity endorsement

While Fairtrade has successfully tapped into certain sectors of the public, celebrities have played a vital role in generating a wider appeal. As opinion leaders, whose thoughts and habits command widespread attention, celebrities can help popularise Fairtrade in uncharted or unfamiliar territory, and can through coverage of their visits overseas help cross the divides between members of the public and farmers overseas. Furthermore, they can generate the sense that others are also buying Fairtrade and so convince people that it is safe and worthwhile.

Harry Hill, a self-styled 'celebrity volunteer' for Fairtrade, had never been to a developing country before visiting cocoa and banana producers in Ghana with the Fairtrade Foundation. His quirky sense of humour has enabled him to talk about Fairtrade in a way that removes it from the 'charity box.' 'By buying Fairtrade products,' he says, 'real people in the [developing] world benefit directly – and how often does that happen?'

A highly effective photo exhibition profiling celebrity Fairtrade supporters was commissioned for Fairtrade Fortnight 2006, showing how celebrities such as Anita Roddick, Gail Porter, Starsailor and Harry Hill, among others, are changing the way they shop and making Fairtrade their habit. 'This is not about a single action people can take,' noted exhibition photographer Trevor Leighton, 'but a small change they can make on an ongoing basis – helping individuals every time they shop.'

Not only do such celebrity statements of support garner substantial publicity for Fairtrade, but they have been critical in establishing Fairtrade's position in mainstream society. Since the early days, the Foundation's newsletter has been fronted by celebrities, including Angus Deayton and Vic Reeves (who appears on the cover of the Spring 2006 edition of the Fair Comment newsletter). Commitment to wide trade justice campaigning from stars such as Coldplay's Chris Martin has also gone hand-in-hand with support for Fairtrade.

Farmers and workers centre-stage

In all these differing activities and promotions, the Foundation has sought to emphasise the individual stories of farmers and workers and to ensure that they are as centre-stage in the marketing of Fairtrade as they are, of course, at the heart of the concept itself. Together with strikingly personal photos, examples of producers' stories appear on posters, leaflets on product packaging and, importantly, on the Foundation's well-visited website.

The message has focused on the most tangible benefits of Fairtrade. Producers tell their stories in direct quotes, focusing on the difference Fairtrade has meant to them, their families and their communities. They have, for example, often spoken of investing the premium in specific projects such as schools, wells and primary clinics. The main focus has been on the Fairtrade pricing mechanism and the resulting concrete community achievements to which the public can readily relate. There have also been attempts, however, to relay the more complex benefits such as market access, producer organisation, selling direct and cutting out local middlemen, right through to the more abstract notions of 'empowerment.'

Qualitative research shows that the public warm to the human touch. They react positively to photos, stories and facts about individuals benefiting directly from Fairtrade. As well as the emotional appeal, such real stories give Fairtrade legitimacy and consumers report gaining reassurance from them. Interestingly, the focus on farmers and workers also gives people reassurance about quality as it stresses the more direct provenance of the food, coming from small family farmers.[10]

The direct involvement of producers represents a unique aspect of Fairtrade marketing. The Foundation facilitates speaking tours of the UK by producers, giving the media an unparalleled opportunity to tell first-hand stories about the real-life impact of Fairtrade. Putting the spotlight on farmers brings alive the anonymous concept of the 'trade' behind the products we all enjoy (see Figure 3.6). The producers likewise benefit because they can personally meet their buyers in companies and meet the consumers, thus better understanding the market demands. For example, in 2006, fourteen producers toured the UK to speak about the benefits that Fairtrade has brought to them and their communities. These included Maria Sergeant, a banana farmer from the Windward Islands, Silver Kasoro-Atwoki, a tea grower from Uganda and Agrocel Cotton Project Manager, Shailesh Patel, from India.

Underlying the focus on producers, has been an analysis of the system of world trade, exposing cases of structural injustices. Fairtrade highlighted, for example, the catastrophic fall in coffee prices in 2002, when the commodity price for coffee hit an all-time low (see Figure 3.7). The Foundation having drawn attention to the impact these prices were having on farmers, the BBC visited Nicaragua in 2003 to document rising levels of malnutrition among coffee communities.[21]

Against an analysis of the structural problems and negative impact of much mainstream global trade, as particularly highlighted by the wider coalition of over 60 NGOs in the Trade Justice Movement, Fairtrade has provided a resolutely positive message underlining the ability of individuals to create change for the better. The overall feel of Fairtrade messaging is thus positive and empowering. It seeks to cut through the prevailing cynicism and convince consumers that their purchases really do make a difference, often juxtaposing photos and words of producers with photos and words of ordinary or celebrity Fairtrade consumers. The message is: 'You don't have to wait for Government to move; you don't even have to wait for companies, because you can push them into acting by buying Fairtrade products or asking for them.'

Figure 3.6   By putting farmers and producers centre stage, Fairtrade emphasises a real human connection behind the products we all enjoy.

Mind the gap

In recent years, while awareness and understanding of the FAIRTRADE Mark has steadily risen to half the population, not all of those people are buying Fairtrade products. Consequently, the Foundation has consistently sought to narrow the

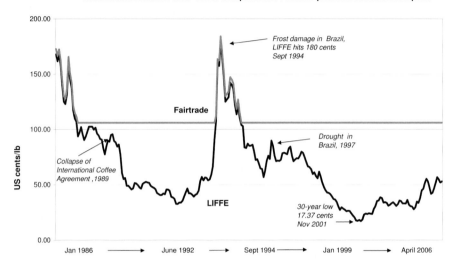

Robusta Coffee Market 1986 - 2006: Comparison of Fairtrade price and London LIFFE price

Figure 3.7   By guaranteeing a fairer deal for producers, Fairtrade allows farmers to benefit despite volatile price fluctuations in markets such as coffee.

margin between people's knowledge and evident support for Fairtrade and their daily purchasing habits. Indeed, narrowing that gap between awareness and stated intentions with actual behaviour is the major challenge for all schemes that seek to assure consumers.

In earlier years, focus group research showed that one key reason behind this gap was a perception among non-purchasers that products were of poor quality. Resistance was not necessarily based on experience but on the assumption of non-buyers that because the products were Fairtrade, with moral and social reasons to buy, quality would somehow be inferior. Consequently, both the Foundation and licensees have worked hard to improve the quality of the products themselves and equally importantly, to communicate the level of this quality. Extensive sampling has been key to overcoming perceptions.

Head of Marketing at Cafédirect, Sylvie Barr, underlines quality as a key concern: 'We have a close relationship with our growers and they understand the importance of quality. They say to us 'tell us what we need to do to improve the quality of our produce.'

Cafédirect's 2004/5 Annual Report states that dozens of young people were trained in 'cupping' and quality control in Mexico and Nicaragua. The report likewise states that during the Nicaraguan 'Cup of Excellence' Coffee Competition, more than half of the winners were Cafédirect partners, with the Prodecoop Cooperative finishing top for the second year in a row.[22] This quality drive is reflected, moreover, through packaging and presentation. For example, the name of Cafédirect's freeze-dried instant coffee, 5065, refers to the height at which the best quality coffee grows, with the tag line 'The height of coffee taste' emphasising this focus while the packaging takes you from the cup to the origin.

With very different fresh fruit products, Fairtrade Development Manager, Geoff Chappell, in an interview with the Fairtrade Foundation said that Malet Azoulay UK has also included quality in its marketing mix: 'Our most important marketing tools have included delivering high quality fruit consistently, developing the Fairtrade range, the use of clearly defined Fairtrade packaging with labelling that incorporates the Fairtrade message and delivers some of the producer stories.'

The Foundation has also in its own materials sought to change perceptions both directly and indirectly, while being careful never to give the impression of being a quality label. Yet all of its materials nonetheless seek to convey an impression of good quality, with enticing product shots. In 2004, the focus on quality was explicit, as a Fairtrade Fortnight poster declared: 'Gabriel gives the pick of his crop to his favourite customers' (Figure 3.8). Accompanying materials detailed how producers do indeed prioritise their best crop for Fairtrade, how they use the premium to invest in quality improvements, and how Fairtrade products have won quality awards.

The Fairtrade Foundation has always used recipes to help entice the public, and celebrity chefs have played a vital role in underscoring quality, as when wine critic Oz Clarke recommended Fairtrade Chilean wine. Similarly, Hugh Fearnley Whittingstall explained how he chooses Fairtrade alongside organic and local produce. In 2005, as part of the Make Poverty History Campaign, the Fairtrade Foundation partnered with Channel Four chef Vicky Bhogal, Oxfam and publishers Simon & Schuster to produce a celebrity cook book: *A Fair Feast*.[23]

The work – on the products and on the marketing side – seems to have paid off, as consumer perceptions about the quality of Fairtrade products have turned right around. Perceptions of lower quality have sunk right down the list of barriers to purchasing. Indeed, the press increasingly focused on the top quality of Fairtrade products, including such favourites as Green and Black's Maya Gold, which has been explicitly sold as a luxurious top quality chocolate.

Focus group research has shown that, when pushed, people often do in fact themselves associate Fairtrade with small farmers and so with equal and often superior quality goods, with less intensive agricultural or manufacturing methods. They feel reassured knowing the food can be traced back to individual farmers whom, they assume, may use fewer chemicals and more natural methods and greater care.

By 2006, it was clear that the key barrier was now overcoming old habits, as outlined above. The focus was therefore on increasing visibility of the products in

Figure 3.8   Posters such as these emphasise the high quality of Fairtrade products and the efforts farmers make to produce quality crops for their Fairtrade customers.

store. So many people warm to the idea but will not search out the products if they are hard to find on busy shop shelves. Hence the focus on getting the Fairtrade habit, backed by unprecedented levels of signage in shops to help consumers find and buy the products. Converting increased awareness into purchases remains the core challenge if more producers in more countries are to benefit from Fairtrade.

The international reach of Fairtrade

The Fairtrade Foundation is part of an international network, the Fairtrade Labelling Organizations International (FLO). Encompassing national initiatives across 22 countries and producers across 58 countries, FLO officially owns the FAIRTRADE Mark, and sets and monitors the international standards behind the FAIRTRADE Mark.

While the certification system is international, with the same FAIRTRADE Mark being used everywhere save in Mexico, Switzerland and the US, marketing strategies are still developed on a national basis. In many cases, the strategies are almost identical: the focus on the producers' personal stories, on quality, working with celebrities, for example, are common themes. Furthermore, effective ideas are being adopted globally such as the Fairtrade Towns movement, which is spreading across

Fairtrade in facts: a quick look at Fairtrade

- Globally, Fairtrade benefits some five million farmers, workers and their families in 58 developing countries
- In 2005, there were 550 producer groups participating in Fairtrade throughout the world, up from 260 certified producers in 2003
- The number of producer groups in Africa selling Fairtrade products to the UK market rose 50% from 2005 to 2006
- Global sales of Fairtrade products hit the $US1bn mark for the first time in 2004
- Worldwide sales of Fairtrade products continue to grow by around 40% each year, and grew by 56% between 2003 and 2004 alone
- Europe, with 15 Fairtrade labelling initiatives and sales figures close to €600m, is the world's most advanced Fairtrade market
- The UK is the world's single largest Fairtrade market, with 40% of British households buying at least one Fairtrade product in 2005
- In 2006, Fairtrade consumers in the UK were purchasing an estimated £220m worth of Fairtrade products from more than 300 producer groups from countries in Africa, Latin America, the Caribbean and Asia

- Since 1999, money spent on ethical food items in the UK (including Fairtrade and organic) has steadily grown to top more than £4bn
- In the UK, by mid 2006 Fairtrade consumers were purchasing more than 1500 Fairtrade certified products from 212 different companies – a leap from 130 products in 2003
- In the UK, an estimated 20% of roast and ground coffee, 5.5% of bananas and 5.0% of tea is Fairtrade certified
- Cafédirect, the UK's leading Fairtrade company, is the fourth largest roast and ground coffee brand in the UK and according to a recent survey by Reader's Digest, ranked an impressive third out of 467 household brands for social responsibility, achieving a higher score than The Body Shop and the Co-operative Bank, amongst others

Europe. In other cases, the differences between national situations – the US versus Finland, for example – are so compelling that diverse strategies must be adopted in order for the Fairtrade movement to progress. The national organisations are nonetheless working together with increasing frequency, looking to collaborate more on marketing initiatives while respecting important national objectives. Undoubtedly the global reach of the system, together with its core involvement of producer and trader representatives as the key stakeholders on the governing board and key committees, are central to Fairtrade's current and long-term success.

The future of Fairtrade

The Foundation's vision is that all products coming from disadvantaged producers in developing countries can in the end become Fairtrade certified. It is working towards 75% recognition of the FAIRTRADE Mark, deeper understanding of Fairtrade, and expects sales to continue their steep upward curve so that more producers in more countries selling more products can participate in Fairtrade. Fairtrade has come a long, long way from when the first products were launched. The strategies of marketing Fairtrade will of course need to be developed for the next stage in its growth. Many core planks will remain, and indeed be strengthened, as the potential for the grassroots movement to continue expanding and strengthening is still enormous. New planks will certainly be added, however, as the ever-growing audience, opportunities and challenges shift for Fairtrade and its vision of a world in which every person, through their work, can sustain their families and communities with dignity.

References

1 Vidal, J. (2003) Retail therapy. *The Guardian*, 26 February.
2 Fairtrade Study (2006) *TNS Omnimas Topline Results*, April. TNS Omnimas, Epsom.
3 Mintel (2004) *Consumers' Ethical Concerns, Attitude Towards Ethical Foods*. Mintel, London, February.
4 Elustondo, M. (2004) *Ethical Consumers Plump for Fairtrade Commodities*. PA Newswire Services, London, 23 April.
5 Armstrong, M. (2005) Survey shows nine out of 10 want ethical feedback. *The Guardian*, 28 November.
6 MORI (Market and Opinion Research International) (2005) *Fairtrade Tracker Study Results*. MORI, London, April.
7 Marks & Spencer Media Centre (2005) *Spotlight on Africa leads to increased UK appetite for Fairtrade products*. Marks & Spencer, London, 4 August.
8 Marks & Spencer Media Centre (2005) *Spotlight on Africa leads to increased UK appetite for Fairtrade products*. Marks & Spencer, London, 4 August.
9 The Co-operative Bank, nef and The Future Foundation, (2005) *The Ethical Consumerism Report 2005*. Manchester, 12 December.
10 Dragon Brand Agency (2005) *Inspire: Consumer Understanding of Fairtrade*. Dragon Brand Agency, London, November.
11 Research Works Limited (2005) *The FAIRTRADE Mark and Fairtrade Fortnight: Presentation of Qualitative Findings*. Research Works Limited, Finchley, October.
12 Research Works Limited (2005) *The Fairtrade Mark and Fairtrade Fortnight: Presentation of Qualitative Findings (October 2005)*. Research Works Limited, Finchley, November.
13 TNS Superpanel Research (2006) TNS Research. www.tns-global.com
14 Mintel (2004) Executive summary. In: *Attitudes Towards Ethical Foods*. Mintel, London.
15 National Consumer Council (NCC) and Sustainable Development Commission (SDC) (2006) *I Will If You Will: Towards Sustainable Consumption*, May. Funded by the Department for Environment, Food and Rural Affairs (DEFRA) and the Department of Trade and Industry (DTI). www.ncc.org.uk/publications/index.htm
16 Pati, A. (2005) Most innovative charity winner: the Fairtrade Foundation. *Third Sector Magazine*, 9 November.
17 Cheliotis, S. (2005) *FAIRTRADE Mark is top of class in Superbrands Awards*. Fairtrade Foundation Press Release, 28 September.
18 The Co-operative Group (2003) *Coffee: What a Difference a Penny Makes*. The Co-operative Group, November.
19 Cafédirect (2005). Online Press Office. http://www.cafedirect.co.uk/pressoffice/release.builder/00042.html, 7 October.
20 TNS Superpanel Data (2006) *UK Organics Consumers – Where, What and Why*. TNS Research, May.
21 Carslaw, N. (2003) *Child Victims of Coffee Trade Wars*. BBC News, 10 February.
22 Cafédirect (2005) *Cafédirect's Annual Report and Accounts 2004/5*, p.6. http://www.cafedirect.co.uk
23 Bhogal, V. (2005) *A Fair Feast: 70 Celebrity Recipes for a Fairer World*. Simon & Schuster, Cambridge.

Successful Organic and Fair Trade Brands

Chapter 4

Case History: Yeo Valley Organic

Graham Keating
Director of Communications, Yeo Valley

The origins of Yeo Valley

The Yeo Valley story began on a small dairy farm in Blagdon, 12 miles south-west of Bristol. Roger and Mary Mead bought the 150-acre Holt Farm in 1961 and Roger, who had been raised in a dairy farming family, set about developing his British Friesian herd. From the very beginning, his skills as a dairyman led him to choose a pedigree herd ideally suited to the mild, wet Somerset climate and his entrepreneurial spirit also gave him a vision to find a way of selling his milk directly to his own customers (Figure 4.1).

Figure 4.1 Roger Mead (second from left) with visitors from the Department of Agriculture of Barbados in the late 1960s. The pre-cast concrete barn was to become the Yeo Valley yogurt dairy.

This vision moved a significant step closer to reality in the early 1970s when the neighbouring, 60-acre Lag Farm became available. Figure 4.2 shows an early photograph of the farm. Though not large, the farm came with useful buildings and was close to the passing traffic on the main road, so the Meads started a Pick Your Own Fruit venture and opened a small tea room for the public. Local-baked scones and home-made jam were offered to the fruit-picking public and the tearoom menu soon featured Roger's first commercially produced dairy product, clotted cream. Of course, taking the cream left the farm with skimmed milk so he used it to make his first trial batches of yogurt, then a fairly new product to many British consumers. So it was that, as a by-product of a farm teashop, Yeo Valley Farms (Production) Ltd was born in 1974.

Production in the early years was on a very small scale, using milk churns placed in pens containing water to warm them in order to incubate the yogurt. The first manufacturing room was a converted barn, deliveries were made in an unrefrigerated Morris Minor van and the first customers were local shops and hotels, but Roger's yogurts found an enthusiastic customer base.

The following years were testing, as they would be for any small food business, but by the early 1980s the dairy employed around 30 people and had attracted business from some of the emerging supermarkets for retailer-branded yogurts, a market that was to become the core of Yeo Valley's business. The Lag Farm yogurt dairy was

Figure 4.2   Lag Farm Dairy in 1976, the barn extended for yogurt production.

expanded to meet the growing demand and Yeo Valley soon won a reputation for quality, flexibility and service that would result in dramatic growth in the 1990s.

The farming focus at Yeo Valley

While this new dairy business demanded huge concentration, Roger didn't neglect his first love, dairy farming. He gradually extended his farm, as neighbouring land became available for rent or purchase, and the Lakemead British Friesian herd expanded. He moved some of the land into arable production to provide the cows with home-grown winter fodder and eventually owned land up on the Mendip plateau some three miles from Blagdon to complement the low-lying Holt Farms land. For ease of management and welfare standards, the herd was eventually split into two milking herds – Holt Farm beside Blagdon Lake and Yoxter Farm on the Mendips – and the relatively prosperous farming times of the early 1990s allowed Roger to invest in high-quality buildings and facilities. He also began the process of re-investing in the environment, starting a process of countryside stewardship, which continues today, on land that had seen years of under-investment under previous ownership.

Today the farming enterprise extends to 1250 acres and there are 420 dairy cows whose milk is delivered daily to the Lag Farm dairy. The farm continues to see investment in environmental standards, with a Yeo Valley conservation team laying hedges, rebuilding traditional dry-stone walls, creating wide, permanent set-aside field margins and planting thousands of native broadleaf trees. The farm is now partially organic – about 20% is farmed organically for sheep and beef – and the lessons learned in this venture are being applied to the remainder of the farm. Overall, the farm is an example of how farms could and should be if only farming finances would allow. Holt Farms now has the crucial advantage of the bigger business that it spawned, Yeo Valley, to help support the cost of maintaining these higher standards.

The farm has remained a key part of the Yeo Valley story and indeed is an important aspect that differentiates the business from its competitors. Supermarket buyers are taken on a tour of the farm as a regular part of their visits to the yogurt dairy, something that has become very unusual in British food manufacturing. Most food processors have become increasingly devolved from the farms that produce their raw materials.

The successful completion of Roger's farming vision was, tragically, not something that he lived to see. He died in a tractor accident in 1990 at a crucial time for both the farm and the growing Yeo Valley dairy business. His widow, Mary, took over the reins of the farm and his son Tim, then aged 27 and recently returned to Somerset as a qualified accountant from the City, was faced with the onerous task of taking over his father's yogurt enterprise. The combination of the family skills, together with those of a small management team already in place, took the company through a turbulent time and it was Tim who led the dairy company into a new venture for which it has now become best known – Organic.

Yeo Valley Organic – the early years, 1994–2000

In 1994, the Milk Marketing Board monopoly was broken up by Government directive and the British dairy market went through a period of intense change. The old buying relationships changed, new buying groups emerged and most of the bigger dairy processors seized the opportunity to sign up dairy farmers to deliver directly to them.

In the midst of all this activity were a handful of pioneering organic dairy farmers such as Sally and Henry Bagenal in Loxton, Somerset. These farmers had turned to organic methods largely out of concern for the environment and health rather than through public demand for organic products. In 1994, organic food was a tiny and much undeveloped market – something that was to change dramatically in the next couple of years – but, in the midst of all the activity in the conventional dairy market, these organic farmers were finding it hard to secure a buyer for their milk.

The Bagenals approached Yeo Valley with a proposition that the business might consider producing an organic yogurt and Tim Mead's response was both encouraging and practical. He pointed out that organic yogurt production would need a reasonable volume of sales to make it viable but also that yogurt only represented a small percentage of milk used in the conventional market. If a new organic dairy market was to prosper it would need to find markets for liquid milk, cheese, butter and cream and the handful of farmers would really need to work together to develop a sustainable supply base. He offered some financial support to explore how co-operation between the farmers might be legally formalised and so the Organic Milk Suppliers Cooperative (OMSCo) was born.

Presented with a challenge to take the hitherto unwanted organic milk and transform it into a commercially viable product, Yeo Valley used its retailer-brand experience to produce a high quality natural yogurt; a pot was hastily designed and shown to existing retail customers (Figure 4.3). Tesco was one of the first to take the new Yeo Valley Organic Natural Yogurt alongside a Tesco-labelled product but other retailers quickly followed suit. For the supermarkets, the launch of organic yogurt was a low risk since it came from an existing supplier, Yeo Valley, which could use its existing economies of scale in packaging, purchasing, transport and manufacturing capabilities to produce a product that could be competitively priced. Since Yeo Valley at that stage had no real marketing costs to worry about, it priced the organic product at the same price as conventional, initially to ensure that each batch would sell to consumers who were then largely ignorant of organic food and farming.

The outbreak of BSE in 1996 provided a huge impetus to the infant organic market as the widely reported food scare motivated consumers to search for apparently less

Figure 4.3   The original Yeo Valley Organic packaging.

risky sources of food. Yeo Valley had by then expanded into making fruited yogurts and responded to the accelerating consumer interest in its products by working closely with OMSCo to encourage new farmers to convert to organic dairying. The growth in demand began to put strains on the capabilities of Yeo Valley's Lag Farm dairy where the added complications of strict segregation of organic and conventional products brought its own inefficiencies. It was time to find a new organic dairy and Yeo Valley also decided, crucially, that it was time to put more, concentrated focus on its new organic business.

A new company was formed, The Yeo Valley Organic Company Ltd, in 1996 and the Dairy Crest Cannington Creamery was purchased in 1997 as the new home for Yeo Valley's organic products. The Cannington site had been Dairy Crest's new product development dairy, claiming the launch of Lymeswold Cheese as part of its somewhat chequered history but, by the mid 1990s, it was in steep decline and was scheduled for closure. Yeo Valley's purchase safeguarded the jobs of the remaining

40 skilled dairy employees, engineers and managers and, of course, gave the new company an excellent base for its growth.

The new Yeo Valley Organic company continued to purchase its commercial, procurement and logistics services from the original Yeo Valley, which was continuing to grow its sales of non-organic, retailer-brand yogurts. Having a bigger, established sister company allowed the new organic business to borrow machines and skills as needed but, crucially, its new identity provided the management team with the opportunity to focus solely on the needs of the still new Yeo Valley Organic brand and on the production facilities required to support it. In truth, the brand marketing activities at this stage amounted to little more than packaging designs for new flavours and some low-key PR stories. Consumers had by now begun to discover the Yeo Valley Organic range in more and more supermarkets and word-of-mouth recommendation was driving dramatic growth in sales. The biggest challenge to the Yeo Valley Organic team at that time was to secure sufficient supplies of organic milk and to find new supplies of organic fruit.

For its liquid milk supplies, Yeo Valley continued to develop its relationship with the growing OMSCo, whose founding Chief Executive, Sally Bagenal, was actively encouraging new members to join. The surplus milk supplies of the early years were now a thing of the past and new supplies were urgently needed. Yeo Valley Organic signed a ground-breaking three-year rolling contract for its milk with OMSCo in 1996, guaranteeing a fair milk price for increasing milk volumes. It also launched a profit-related bonus to OMSCo members, which, in 1999, paid £103 000 to the farmers. This bonus was later transferred into an increased milk price but it highlighted Yeo Valley's commitment to a fair and equitable relationship with its organic dairy farmers at a time where the future market for its products was still by no means secured. Through joint negotiation with OMSCo and Yeo Valley, Sainsbury's signed up to a long-term contract for organic milk with OMSCo in 1999, an act that finally brought a level of confidence to many more farmers to convert.

In May 1999, Tim Mead appointed a new Managing Director to the new business to lead the installation of new production facilities at the Cannington dairy and to seriously develop the marketing activities of the Yeo Valley Organic brand. This resulted in the first redesign of the original, rather 'hand-drawn', packaging and the recruitment of the brand's first, and current, Marketing Director. It was time for the Yeo Valley Organic brand to stand up and be counted.

Development of the brand 2000–2002

By the end of the 1990s the brand was now established with a small but loyal consumer base, yet it was still relatively unknown to the wider yogurt-buying public. Its on-shelf presence had developed through the creation of new flavour additions to its range rather than through a classical brand strategy approach. The company's focus on assisting the development of the UK organic milk supply chain had also influenced the brand range; Yeo Valley Organic was purchasing whole milk from its farmers so had to use all the components of the milk. Wholemilk and fat-free yogurt production left cream unused so the Yeo Valley Organic range developed into products such as butter, crème fraîche and retail cream at an early stage to use the 'whole cow' to minimise inefficiency costs (Figure 4.4).

To grow the brand beyond the niche it had created for itself, it was clear that Yeo Valley Organic needed to appeal to a wider consumer audience, consisting of people who had yet to be converted to organic food. At the same time the brand's main customers, the multiple retailers, were beginning to expect more from Yeo Valley in the way of brand advertising and support; with a surge in general consumer awareness of organic there was suddenly a number of new organic brands – or organic variants of existing non-organic brands – so the market had become competitive. The UK saw the arrival of brands such as The Enjoy Organic Company and Seeds of Change as well as the appearance of organic versions of long-established brands such as Heinz. The supermarkets were also pushing out organic own-label products across all their product categories as quickly as they could.

Yeo Valley responded with its first ever national media advertising campaign, supported with much more focused PR activity in the consumer press. The process of developing the adverts highlighted something of a difficulty: the team felt there were so many interesting and compelling stories to tell but consumers, when asked, seemed to want a very simple message centred on apparently very selfish themes of taste and value for money. The core organic consumer, labelled the 'dark greens' by the

Figure 4.4   The Yeo Valley Organic range in 2000.

marketers, wanted to be told all the details of organic farming practices, ingredients lists and packaging information while the new consumers that the brand was hoping to attract, the 'dabblers' in organic, found such information to be of little relevance in their purchasing decisions. For the Yeo Valley Organic team, with aspirations to tell a real, detailed and, they hoped, engaging story about great farmers and wonder-ful-tasting products, it was somewhat dispiriting to watch consumer focus groups dismiss their proposals as being just 'too much information'. 'Does it taste nice, is it good for me and will it make my hair shiny?' seemed to be a common outcome of the relatively few research groups that the brand conducted at this stage.

The first press and poster adverts of 2000 were, as a result, rather lighter in tone and content than had originally been envisaged (Figure 4.5). The Yeo Valley Organic cartoon cow became the brand's advertising image, in spite of the team's

Figure 4.5   One of the first Yeo Valley Organic poster and press adverts, 2000.

initial and firm assertion to the creative team that 'we don't want a cartoon cow for our brand!' However, the adverts showed a lightness of tone and a touch of humour that had not been seen in organic products before – though, in truth, there had been very little advertising of any kind for organic products up to then. The Yeo Valley Organic cow appeared in a variety of guises in consumer magazines and on posters and generally raised a smile. The adverts appealed to the non-organic consumer as a friendly, approachable aspect of organic food though some of the themes raised a few eyebrows: the 'There are no fat cows at Yeo Valley Organic' poster placed in a national chain of gyms and health spas caused some debate.

This first foray into advertising ran for over a year but the company's available funds to support such activity were limited. There was a pressing need for investment at the Cannington dairy for new capacity to keep up with the steadily growing demand for the brand. Meanwhile, there was a realisation that the marketing story needed greater depth to make the Yeo Valley Organic brand stand out against a whole range of new organic brands. Such was the rush of new organic products to market, the concern at Yeo Valley was that the organic message was in danger of being devalued and, potentially, tarnished.

It was clear from Yeo Valley's own data on yogurt, that retailer brand organic products were not succeeding in establishing anything like the market share of their non-organic cousins. With a slowly-growing non-organic yogurt market having over 30% of its sales through retailer brands, the figure for the organic market was less than 16% in 2000 and these sales were achieved at a cost of a large number of new product launches and relatively hasty delists. The launch of the Marks & Spencer organic yogurt range in 2001 illustrated the peculiar difficulty of the organic market place; within a year the range would be withdrawn because of lack of sales, not to return for another two or three years. Indeed, except in sectors such as fruit and vegetables, retailer brand organic did not seem to be working in any area where there were organic branded alternatives.

Meanwhile the 'new organic brigade' of brands – often produced by large, hitherto non-organic food companies – together with organic versions of existing, well-established brands were bringing new marketing activity to the organic sector, yet the failure rate was surprisingly high. In Yeo Valley's own market, dominant non-organic brand Müller finally launched an organic variant in 2001, supported with significant advertising and promotional activity, yet it too was withdrawn after less than a year. Yeo Valley identified the emergence of three distinct types of organic brand, illustrated in a Yeo Valley's presentation given to the Soil Association national conference in January 2002 (see Figure 4.6).

History, of course, has shown that the 'true' organic brands are still very much in existence today and are thriving in their relative market sectors; the 'new' organic brands have fared less well (RHM's Enjoy Organic brand was quietly removed from the market in 2004 whilst Mars' Seeds of Change brand remains) and many of the 'Portfolio' organic products have disappeared.

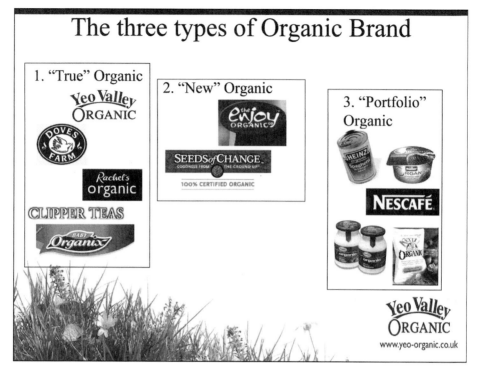

Figure 4.6   A slide from the Yeo Valley Organic presentation to the Soil Association National Conference, January 2002.

The common reasons for many of the failures centred around, first, a lack of true organic credentials and, second, packaging design. The heaviest purchasers of organics, the dark greens, were, by their nature, very savvy consumers and, for new products to succeed, were the group that had to be convinced as they bought such a high proportion of total organic sales. For 2002–2003, 83.7% of organic sales were purchased by the dark greens, who accounted for only 23.1% of organic purchasers. They were then, and remain today, wary of marketing stories above substance, so took some convincing to adopt any new organic brand and they largely rejected organic versions of well-known, non-organic brands as being motivated by somewhat cynical commercialism. Similarly, they felt that retailer-brand organic products gave them little confidence that the products embodied the full organic ethos. The dabblers, meanwhile, needed to be enticed to sample new products but were fairly infrequent and financially insignificant purchasers of organics. Unfortunately for them, many of the new retailer-brand and new brand products adopted a rather dull, earthy look to their packaging which rather re-enforced many consumers' pre-conception that organic food might be good for you but probably tasted like wholegrain muesli.

Yeo Valley's response to the new and busy organic sales fixture was to go back to the drawing board with an exercise to clearly define the brand's true essence (Figure 4.7).

This 'map' of the brand formed the basis of further work on the development of the product packaging. The move to the very uncluttered but rather stark packaging in 1999 was softened with the introduction of the Yeo Valley Organic grass theme that had been a part of the original branding back in 1994. As confidence in the brand developed a more radical change to the design would take place in 1993 with the development of the current 'tick' theme and the further iteration of the logo.

The evolution of the brand logo (see Figure 4.8) also illustrated the growing confidence in the Yeo Valley name as the key deliverer of a quality message to the consumer. The word organic remains an intrinsic part of the logo but was gradually reduced to a supporting rather than a dominating role.

During this era of the young company's short history, Yeo Valley began to form its own opinions on the organic market as a whole. Over the late 1990s and early 2000s, while the market was in a period of growth and change, Yeo Valley had started to express strong views on the sustainable development of the organic market. These opinions were shared (whether they were asked for or not) with retailers, the Soil Association and other organic processors. The company had written into its Memorandum & Articles of Association (the governing rules for its directors) a set

Figure 4.7   The Yeo Valley Organic range in 2002.

1994 logo   2000 logo   2003 logo

Figure 4.8   The evolution of the Yeo Valley Organic brand logo, 1994 to present day.

of clear objectives for itself, which it shared widely, encouraging others to adopt something similar (see box below).

As the new millennium came, Yeo Valley Organic increasingly began to articulate its concern about the way the UK organic market was developing in terms of scale and true 'organicness'.

At a presentation given at the Marketing Week Organic Conference in London in June 2001, Yeo Valley Organic identified three types of organic food manufacturers, and superimposed them onto graphical axes of size versus 'Organic vs. Conventional beliefs' (see Figure 4.9).

Yeo Valley predicted that the 'organic founders' would remain 'small and content with their limited production capacity, few systems, strong beliefs in minimising food miles and no aspirations to supply the supermarkets.' Their consumer would find their products at the farm shop and farmers' market and, today, many of them continue to thrive there.

The 'organic entrepreneurs', it was suggested, would make 'niche-products in small scale but would be focused on quality and innovation with good but basic operating systems. They might supply the supermarkets on a regional basis but their

---

The Yeo Valley Organic Company's objectives are:

- To promote the primary production of organic food from environmentally friendly farming techniques and good animal health and welfare practices
- To produce healthy organic food using those primary products
- To promote the widespread consumption of organic products
- To educate consumers of the benefits of consuming organic food
  And, as a result, to improve:
- The overall environment and human health.
  In seeking to achieve these objectives the Company intends to deal fairly and equitably with its suppliers and customers in all its business dealings to ensure that its suppliers receive a fair return and that its customers only pay a fair price.

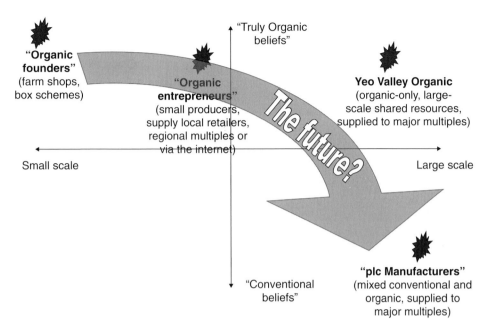

Figure 4.9   Slide from the Yeo Valley Organic presentation at the Marketing Week conference, June 2001.

business growth would be limited, not by consumer demand but by the availability of investment capital and the ability to attract the required calibre of people to their companies. They would be vulnerable to product delists, new market entrants and take-overs.' Today many such entrepreneurial companies and brands are now owned by multinationals.

The 'plc manufacturers' would, by their nature, be 'large-scale, mixing conventional and organic production with good production systems'. Yeo Valley suggested, however, that they would be 'far-removed from the organic story; they will purchase materials from wherever they can. There may be conflict of organic against their conventional products, especially if branded; some may have a vested interest for organic to fail'.

The large arrow labelled 'The future?' on the graph highlighted that conventional economies of scale arguments would suggest that as the organic market grows, more and more organic products would move to the plc manufacturers as the only ones with the necessary skills, capital and capacity. Yeo Valley, however, identified a range of risks to the organic sector in the predicted shift towards these kinds of companies:

- Reduced 'organic-focus' in the supply chain due to dual-manufacturing processes
- Decreasing farmer/grower involvement or development

- Cost and margin decisions driven by short-term shareholder pressure
- Pressure may be pushed back to farmers on price and specification so restricting the numbers that might convert to organic farming
- Quality may suffer or corners may be cut in the supply chain
- Consumers will become increasingly wary that 'big-business' has overtaken the organic world
- The credibility of the organic label could become devalued

It's an argument that is still heard today in the organic world, with an additional phrase, not widely used then, of 'food miles' to neatly illustrate some of the pitfalls of industrialised organic production.

Having outlined these threats at the conference, Yeo Valley Organic was proposed as an alternative direction to 'The future' arrow, as a company focused only on organic products and the development of its supply chain, but with the ability to share larger-scale resources and economies of scale with its sister company to enable it to supply the multiple retailers.

Yeo Valley was, and still is, clearly concerned with potential consumer perception that, as a relatively successful consumer brand, it might be thought of by some consumers as being no different to the plcs. The arrival, in 2001, of the Queen's Award for Enterprise in the new category of Sustainable Development, recognising Yeo Valley's key role in developing a sustainable supply of UK organic milk, followed by the BBC Radio Four Food and Farming Award for Best Food Producer 18 months later, was therefore timely and welcome as the company embarked on its next phase of brand marketing.

Developing the brand range and identity, 2002–2004

The close relationship between Yeo Valley Organic and its farmers resulted, in early 2002, in a review of the range of products that the brand offered to take account of the worrying issue of an excess of organic milk from British farmers. After years of a UK organic milk shortage, many new farmers had completed their farm conversions just as predictions for consumer demand, from both retailers and processors, proved to have been very optimistic. Yeo Valley and OMSCo shared the belief that it was vital for this short-term 'blip' to be dealt with as quickly as possible to prevent farmers, many of whom had turned to organic as a last chance of profitable dairy farming, leaving the market altogether.

Yeo Valley's approach to the problem of excess supply was first to maintain its commitment to OMSCo and its long-term contracts. Second, a range of new milk-based products was launched with the Yeo Valley Organic brand now extending to liquid milk and cheese, both product sectors where hitherto most products were retailer branded. Finally, new yogurt and ice cream products were launched to further broaden the appeal of the brand – 2002 was to be a very busy year for the business, in terms of new product launches, as a presentation slide (see Figure 4.10) illustrates.

The arrival of Yeo Valley Organic in the milk fixture, in particular, was important at a time when many retailers' preferred approach to the problem of excess milk was to reduce their retail prices at a time when non-organic milk prices were continuing to increase. The Yeo Valley Organic-branded milk secured a new, higher price point and brought new consumers to the organic milk fixture, so helping increase overall demand for organic milk. It was significant that there was no tangible leadership or direction from the government, the NFU or the other milk co-operatives at this crucial and difficult time for British organic dairy farmers; it was largely left to OMSCo and Yeo Valley to try to stabilise the market through mutually supportive activities as best they could. For Yeo Valley this came at a price – their support for the milk price meant that they were paying up to 4p per litre more than their competitors who were buying on non-contractual terms; so, over the three years from 2002 to 2004, they paid at least an additional £2m per annum over the prevailing market price for their milk. It was this support, however, that was instrumental in keeping many organic farmers in business.

Figure 4.10   A Yeo Valley presentation slide outlining brand launches in 2002.

By late 2002, the brand team had also developed a plan to take the Yeo Valley Organic brand further into the mainstream dairy market in order to grow its overall share. It was clear that the brand had to claim a position of being an accessible but premium brand, through its look and feel, and that Yeo Valley Organic had to decide where it was to compete in the market.

The marketing team reassessed the packaging design, searching for a look that clearly differentiated Yeo Valley Organic on a crowded supermarket shelf. The Rachel's brand had its trademark black pots, which certainly stood out from the crowd but suggested to some consumers that they were an expensive, premium product. Meanwhile images of grass were now being used on a number of different organic and non-organic products so the Yeo Valley Organic packaging no longer had a truly distinctive character. This was illustrated by having the designers mock-up one of the current Yeo Valley Organic pots but with the logo replaced with a Tesco one and it was very evident that the packaging design was eminently transferable to any other brand. This was enough to convince the Yeo Valley team to embark on another time-consuming and expensive redesign, this time with a much greater understanding of the brand ethos and consumer perception to help, or complicate, the process.

A Bath-based design agency was commissioned to develop a new look for the brand. This had to be one that maintained the links to the relatively short brand heritage but that would attract new consumers and provide an unmistakeable Yeo Valley Organic identity. The result was the current design, featuring the 'tick' of grass blades that has been applied across the whole product range (see Figure 4.11).

The new pots were rolled out to the retail trade and into the shops in the spring of 2003 and received a very positive reception.

Figure 4.11   The Yeo Valley tick design applied to the Natural Yogurt range.

A new advertising campaign was designed to support the new products as well as driving consumer awareness and trial of existing Yeo Valley Organic products (Figure 4.12). It was also carefully managed as a project to try to measure the relative effectiveness of different advertising media against other activities such as product sampling. Sales data were purchased for key supermarket stores in the trial south-east test area and a complex matrix of sales results for key products before, during and after the three month campaign was produced, to compare against similar-size stores in non-advertised areas. Additional dimensions to these results were added, so that stores where sampling campaigns of Yeo Valley Organic yogurts had been provided could be compared against stores that had had only advertising or, alternatively, had seen both advertising and sampling.

For people living in the centre and greater south-west London, the Yeo Valley Organic marketing campaign arrived with something of an impact (Figure 4.13).

Figure 4.12   2004 press advert.

Figure 4.13   London advertising campaign, 2004.

Large and small posters, magazine adverts, local paper inserts and vouchers posters on the sides and back of London buses meant that the brand couldn't really be missed. The theme of the adverts was also a long way removed from most consumers' expectations of an organic food brand.

The strap-line, 'In a world gone mad, it's sanely delicious' seemed to strike a chord with many consumers disillusioned with food brands that had proved to deliver rather less than they sold. The 'Food Scientists R not us' advert in particular received a very positive response, though not from the Institute of Food Scientists!

New kids on the block, 2005

Successful though the advertising campaign was, it was clear that getting consumers to taste the products for themselves, through money-off coupons or face-to-face sampling activities, remained an incredibly powerful weapon to gain consumer awareness and, as a result, repeat sales. Many people bought the yogurts for their taste alone, rather than for their organic status, and, by early 2005 there were new, external factors influencing many people's purchasing decisions.

The quality of children's food was to become one of the big issues of 2005, thanks to a certain celebrity chef's TV campaign, and it is widely acknowledged this delivered extra impetus to an organic market that had seen a slowing of its growth rate. Almost 12 months before *Jamie's School Dinners* woke a nation from its slumber on the issue of kids' food, Yeo Valley Organic had researched the children's yogurt market with the help of Campden Food Services and found some worrying results for what most parents would believe were 'healthy' products: of the 18 top-selling children's fruit yogurts in the UK, one-third contained no real fruit at all (just artificial flavouring); those that did contain fruit had only 4.5% on average; the average sugar content was 13.3% and the worst offender contained 21.6% sugar.

Yeo Valley relaunched its two children's four-pack yogurts in the spring of 2005 (Figure 4.14), with the new products containing 50% more fruit (to 9%, or roughly twice the market average) and 10% less sugar than the products they replaced.

The media coverage that these new products received in the parenting press, together with the 'Jamie effect' later in the year, helped drive the sales up despite an increased retail price. Consumers were at last, it seemed, willing to re-evaluate how much they would spend on food if a brand would only give them honest information.

On top of its new children's range and the wider availability of its milk and cheese, for Yeo Valley Organic, 2005 was also the year of meeting the consumer. A new sampling trailer toured supermarket car parks and consumer shows throughout the summer, introducing the range to new tasters. Consumer reaction was very positive and events such as the Glastonbury Festival provided opportunities to develop dialogue with existing consumers and to meet new ones. The combination of a history of steadily building brand-awareness, press and poster advertising, new product

Figure 4.14   One of the new Yeo Valley Organic children's yogurt packs, 2005.

development and direct interaction with consumers, through the sampling shows as well as through the brand's website and direct email enquiries, saw sales in Yeo Valley Organic's core area of yogurt grow by around 19% in 2005.

Such growth was being reflected across the organic dairy sector and, to the relief of hard-pressed organic dairy farmers, the excess of UK organic milk reduced and, by late 2005, had moved towards to a shortfall (Figure 4.15). The challenge to the organic dairy industry, to which Yeo Valley and OMSCo now turned, was one of encouraging new farmers to convert without a repeat of the huge surge in production that had caused the excess in supply in 2002. The demand for organic milk had at last taken it from niche towards mainstream, despite the lack of direction from DEFRA and the unhelpful competition between rival milk groups for whom organic was something of a distraction from daily business. Yeo Valley had accepted a wider role for itself to try to influence retailer strategic decisions, inform decision-makers of the options available and to smooth the bumps and dips of the supply–demand curve, leading by example wherever it could. This strategy, and the close relationship with OMSCo and other regional milk processors, continues to this day.

2006 and beyond

Bringing the history up to date, it will be of no surprise that the Yeo Valley Organic team does not plan to deviate significantly from what has become a successful and

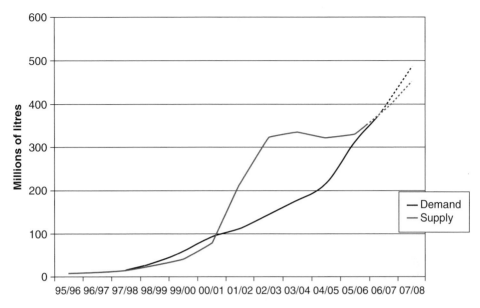

Figure 4.15   UK organic milk supply and demand. Source: The Organic Milk Marketing Report, January 2006; OMSCo.

effective mix of marketing activities, backed by a real and engaging story. With its spring 2006 launch of range of yogurts, called Inner Balance (Figure 4.16), with added seeds and grains to target the fast-growing health market, the product illustrates more of the real Yeo Valley story in its packaging with, for the first time, photographs of the actual Yeo Valley at Blagdon. The separate Yeo Valley Organic Company Ltd was, in June 2004, brought back into Yeo Valley Farms Production Ltd, the original family-run company that spawned it eight years previously; the Yeo Valley Organic brand remains distinct and independent within the bigger business, which has now become the second largest yogurt producer in the UK. This combined size helps the business compete in a market dominated by strong retailer buying power and the increasing need for innovation and points of difference.

Yeo Valley Organic has become recognised by most retailers as an 'anchor brand' in the organic fixture, which attracts knowledgeable organic consumers and first-time trialists alike. As the brand reaches a wider audience, the Yeo Valley team is acutely aware that it must not lose its real identity and difference; many consumers, often with every justification, suspect that successful businesses have sold out, either physically or ideologically, in the quest to achieve commercial success. Just as the first award in 2001 helped reassure the doubters then, the second Queen's Award for Enterprise for Sustainable Development for the whole Yeo Valley business, announced on 21 April 2006, should do much to answer any critics. Today Yeo Valley remains committed to British farming, to the continued development of a growing, profitable and sustainable organic dairy supply-chain and to producing the best quality organic products for British consumers.

Figure 4.16   Inner Balance range, launched 2006.

Figures 4.17 and 4.18 illustrate the size of the Yeo Valley Organic share of the UK yogurt market and the growth of the total Yeo Valley Organic brand in UK yogurt against its competitors over the last five years.

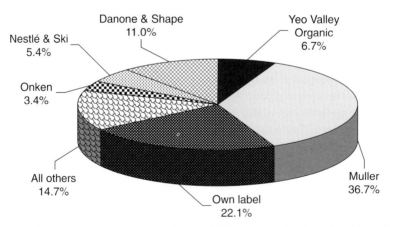

Figure 4.17   The UK yogurt market, by value/total UK yogurt market by value 12 weeks up to week ending 3 September 2005. Source: ACNielsen Scantrack Total GB.

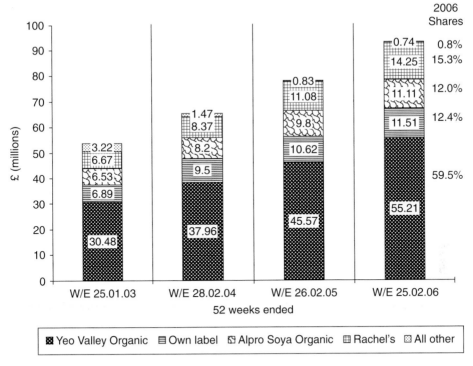

Figure 4.18   The growth of the Yeo Valley Organic Brand. Source: ACNielsen Scantrack Total GB.

The challenge now for the organic dairy industry is how to manage a growth in the market share for liquid organic milk to the 10% already achieved in the yogurt sector while further developing the organic yogurt share. Such growth would bring the demand from organic farms to around 1bn litres per year (around double the current supply) with a retail sales value approaching the £1.3bn currently earned by *all* organic food products in the UK. Yeo Valley Organic's view is that such sector growth would be good news for the beleaguered dairy sector as a whole, injecting some much needed increases in the value of the whole (organic and non-organic) category after years of deflationary forces forcing untenable financial pressures back to farmers.

Conclusions

Yeo Valley attributes much of its success to its farming roots and its continuing independence, coupled with a determination to make products that have a quality and taste which appeal to *any* consumer, whether motivated by the organic label or not.

Such has been the rate of growth, from a very small base, of the organic dairy sector, that Yeo Valley has played an important part in nurturing a sustainable organic supply chain. Such detailed involvement by a processor in the development of its primary farming suppliers would be unheard of in the non-organic world and has demanded significant time and energy from the Yeo Valley team. The growth of the brand would not have been achieved without this effort, however, and it has served to cement a strong working relationship with the OMSCo farmers. This in turn can be communicated as a reassuring story to consumers increasingly concerned about the source of their food.

Yeo Valley Organic has become a key dairy brand with increasing consumer awareness and loyalty. The keys to its success are its authenticity, honesty and quality and these will continue to remain the foundations of its future growth.

Chapter 5

Case History: Green & Black's

Craig Sams
President, Green & Black's Chocolate
Chairman, Soil Association

Whole Earth Foods

The development of Green & Black's owes a great deal to the foundation on which it was built – Whole Earth Foods Ltd. This company, founded in 1967, was dedicated to supplying organic and natural food products to the independent retail trade and to supermarkets. It provided the experience and market knowledge, the grasp of the ethical principles that infuse the organic and wholefood market, the financial and management infrastructure and the business model based on outsourcing and brand management that underpinned the development of the Green & Black's brand. A brief history of the Whole Earth brand is therefore warranted – Green & Black's did not spring into the world as a start-up and it would be a mistake to conclude that its progress happened in a vacuum.

In 1967, 24 years before the launch of Green & Black's, Craig and Gregory Sams started Seed Restaurant, the legendary hip – and hippie – macrobiotic watering hole of the late 1960s, where brown rice and organic vegetables formed the backbone of the menu. Despite today's dour and ascetic image of macrobiotics, the restaurant rocked, both with progressive music and a groovy clientele drawn from the alternative scene of music, the arts and fashion. It was driven by the belief that a mostly vegetarian diet based on wholegrain and organic foods and free of additives was the essential foundation for a sustainable future in a world running out of resources, with a growing population and increasing degenerative disease. Macrobiotics was the driving force of the natural foods industry in the 1960s and the 1970s and macrobiotic principles were at the heart of the evolution of the natural foods diet. Macrobiotic specialties provided a unique attractor that drew regular customers to natural foods stores. The 'no added sugar' boom of the 1980s also derived from the macrobiotic proscription of sugar. Apple juice, however high its glucose or sucrose content might have been, was not called 'sugar' and was therefore an acceptable ingredient in jams and soft drinks.

The success of Seed Restaurant led to the opening of Ceres Grain Shop, Britain's first natural foods store, on the Portobello Road, in the then heartland of alternative society. Soon there were other budding natural foods retailers in search of supplies, forming the customer base for Harmony Foods. Harmony were known as The Brown Rice Barons, because if a retailer bought brown rice in 1970s in Britain it had been imported by Harmony. Harmony Foods made peanut butter under both its Harmony and Whole Earth brands and in 1977 created fruit juice-sweetened jams that launched the whole 'no sugar added' category.

The business thrived on innovation or, at least, innovation provided the finger-nails by which Whole Earth held on as they surrendered territory after territory. It was not enough to be the first with brown rice, miso, wholemeal sourdough breads, brewed soya sauces such as shoyu and tamari, aduki beans, natural peanut butter, no sugar-added jams, organic baked beans and carbonated fruit juice drinks. It was not enough to have led the way in adding fruit juice concentrates in place of sugar in recipes, creating the 'no sugar added' category. There was always someone bigger and stronger who waited until an innovative line got big enough, then did whatever it took to capture the market. Whitworths and Tilda captured the brown rice market, RHM captured the wholemeal flour and wholemeal bread market, Weetabix Alpen captured the muesli market and Robertson's and the (French government-subsidised) St. Dalfour captured the no added sugar jams markets.

Nonetheless, by the late 1980s Whole Earth peanut butter had become a multimil-lion pound brand. It had a respectable 20% of the branded peanut butter market, second to Nestlé's Sun Pat brand. In 1989 Nestlé launched Wholenut – the label artwork had the earthy feel of Whole Earth, the name sounded similar to Whole Earth and the nuts were unblanched, just like in Whole Earth. Nestlé launched a £5m advertising campaign, determined to drive Whole Earth out of the category and capture the 'healthy' part of the market for itself. Whole Earth fought back with a £20 000 spend on tube cards on the London Underground and managed to keep their supermarket listings. Wholenut wasn't finally withdrawn from the market until 1994, but in the meantime it convinced Whole Earth that they needed to broaden and deepen their position.

In 1988 the first organic peanuts had become available and Whole Earth launched the first organic peanut butter. They rushed to capture the niche – with Nestlé, KP and Skippy all pushing into peanut butter, any niche offered some hope of survival. Tesco stocked Whole Earth organic peanut butter in 1990 and the product sold well, merchandised on the main fixture alongside the other peanut butters, long before they created a separate section for organic foods.

The birth of Green & Black's

In late 1990, when a shipment of organic peanuts failed aflatoxin testing, it took seven weeks for a replacement container from Paraguay to arrive. The resulting out of stock situation tested Tesco's patience – they were wasting shelf space on a product Whole Earth were unable to supply. The product was delisted. Whole Earth started looking for a more reliable supplier of organic peanuts. This quest led, in March 1991, to a French aid project in Togo, West Africa, where the agronomist André Deberdt had been working with Ewé tribespeople in the highlands of south-west Togo to grow organic certified pineapple, mango, guava, oil palm, peanuts and cocoa. A sample of peanuts was received but they tested at unacceptably high levels of aflatoxin and Deberdt was notified that they could not be used. He mentioned that the farmers also grew cacao and he offered to send a sample of chocolate made by a French laboratory that specialised in small-scale manufacture. Whole Earth asked for a 70% cocoa solids recipe and a sample duly arrived.

Josephine Fairley was a journalist who'd written several magazine articles about chocolate. She tasted the chocolate and exclaimed, 'This is the best chocolate I've ever tasted. You've got to market it!' Craig Sams had reservations as he was proud to have never sold any foods containing sugar or additives in his career to date. Fairley had £20 000 in the bank from the sale of her flat in Fulham. There was no chocolate on the market that had more than 50% cocoa solids, nothing that was organic and nothing that tasted as good so her money went to purchase the first shipment of the new product.

The biggest challenge was to think of a name. Whole Earth had proudly described itself as a brand that had never sold sugar in anything since 1967, it was a no-sugar brand and had to stay that way. Names like Ecochoc, Organichoc or Bio Choc were considered and rejected as too worthy or ponderous.

What was needed was something that sounded like it had a glorious confectionery heritage, a brand like Barker & Dobson, Charbonnel & Walker or Callard & Bowser, something that sounded like it might have been quietly trading on Bond Street or Jermyn Street since the 1880s. The chocolate was 'green' because it was organic and 'black' because it was the darkest chocolate on the market, with 70% cocoa solids compared to the 49% of Menier or Lindt. 'Green & Black's' had the right ring. After problems enunciating 'Whole Earth,' even to English-speaking listeners, it was good to have something simple to pronounce in any language. A computer mock-up of the design was done in a few minutes and printed out on a colour printer. The design was then converted into artwork and forms the basis of the design that has persisted every since (see Figure 5.1).

The natural food trade was resistant. Sugar was a big no-no in the trade and it's impossible to sweeten chocolate with apple juice, the main sweetener used to replace sugar in Whole Earth's 'no added sugar' products. Community Foods, the biggest wholesaler, refused point blank to stock Green & Black's because it contained sugar. They were adhering to a rigid policy that Whole Earth had done so much to introduce in the 1970s to safeguard the integrity of the natural foods trade. Because they were also a master distributor for the Whole Earth range, however, they were persuaded to reconsider their no-sugar policy. Once they understood that it would compromise Whole Earth's commitment to that arrangement, they somewhat reluctantly stocked the chocolate, thereby making it available to natural food shops. Other natural foods distributors followed their lead.

Figure 5.1   Original pack design.

Sainsbury's were less worried about sugar content and samples were submitted to their confectionery buyer. A few days later Lord and Lady Sainsbury were at a dinner party where a bar of Green & Black's was passed around (see Figure 5.2). The next day Lord Sainsbury asked this confectionery buyer if he'd heard of this new organic chocolate and the buyer could honestly say that it was under consideration. It was soon listed in 12 stores on a test basis. It sold well and not long afterwards it was extended to 20, then 70 stores. Safeway, organic flagship of that era, also listed

Figure 5.2   Original dark packaging – launched in 1991.

it. For Safeway it was almost an automatic listing, they were the major sponsors of Soil Association events and the leading supermarket at the time in terms of depth and breadth of organic offering.

From the outset, Green & Black's communications to the press and consumers focused on the ethical issues, emphasising the benefits to the African producers and the environmental advantages of organic cocoa farming, drawing on the fact that, after cotton, cocoa is the most heavily sprayed crop on the planet.

In 1992, Green & Black's won the Ethical Consumer Award – the first time that the Ethical Consumer Association had granted this award to any company. Vitally important support came from the Women's Environmental Network (WEN), who had just published *Chocolate Unwrapped* by Cat Cox, a book which highlighted the plight of women on large cocoa plantations. The membership of both these organisations were highly motivated activists and felt empowered that there was a product that addressed the issues that concerned them and fulfilled expectations that other companies had dismissed as either idealistic or unrealistic. Bernadette Vallely, founder of WEN, made propagating our message a personal crusade and her activist membership network raised awareness among environmentally conscious women. It addressed the main issues that had been raised in *Chocolate Unwrapped*. So much so that, when the first edition sold out, Josephine Fairley underwrote the cost of publishing a second edition.

Paying fair and guaranteed prices helped the community of producers to plan for the future with more confidence. Moreover, the growers were not exposed to dangerous chemicals such as DDT, malathion and lindane, many of which had already been banned for use in the EU or the USA. A headline in *The Independent* summed it up: 'Right-on, and it tastes good, too' (Figure 5.3).

In 1993, Deberdt's company – the French intermediary link – went bankrupt. Whole Earth bailed him out but it stretched resources to breaking point. Whole Earth now had to finance the import of the beans, the cost of processing and of packaging materials. This required a lot more cash than just buying a finished product. It was unthinkable to discontinue what had become a fast-growing but cash-hungry item, so a way had to be found to finance it. Duerr's, who had been appointed in 1989 to manufacture Whole Earth peanut butters and jams under an exclusivity contract, helped solve the problem. Under that contract all sales of private label peanut butter or no sugar added jams to supermarkets were subject to royalty payments to Whole Earth and this was a sector that was expanding as they captured this business from Nestlé Sun Pat. The agreement was revised to permanently free them from the requirement to make any future royalty payments on their own label sales of peanut butter to supermarkets. In effect they capitalised the value of the future royalty payment stream at a discount. They paid £350 000 to have the contract altered in this respect and the money was used to finance Andre Deberdt's activities while he went through the one-year process of administrative receivership under French bankruptcy law. One valuable asset had been cannibalised to support another, sacrificing longer-term profitability for short-term liquidity, a familiar story to every undercapitalised enterprise.

THE INDEPENDENT

Right-on, and it tastes good, too

ORGANIC CHOCOLATE goes on sale in natural food stores throughout Britain for the first time this month. Green and Black's Organic Chocolate comes from plantations in Togo, West Africa, which have been kept completely free from pesticides and chemical fertilisers, and are certified as organic by the French equivalent of our Soil Association, Nature et Progres.

Sceptics who might assume that any such chocolate, coming from outside established chocolate channels, would be "right-on" but fairly gruesome on the palate, are in for a surprise. The cocoa beans are imported to France, where they are manufactured into chocolate by specialist chocolatiers Soboccum at Dijon, and Pelletier, in St Etienne. The blend of ecological agriculture and French chocolate expertise has produced a winner. Green and Black is a powerful, military-style chocolate with a potent, dark, almost coffee flavour and weights in at a very serious 70 per cent proportion of cocoa solids. Sugar is kept well in the background, making this a sure-fire hit with lovers of long, dark chocolate.

> *Green and Black is a powerful, military-style chocolate with a potent, almost coffee flavour*

Its arrival intensifies the debate around pesticide use in cocoa plantations. A new book, *The Pesticide Handbook* (Hurst, Hay, and Dudley, £22.50 Journeyman), uses the cocoa industry as a case history illustrating the problems of pesticide use in developing countries. One of the authors, Dr Alastair Hay, describes plantations he has visited in Brazil where he found peasant farmers using toxic chemical sprays in choking conditions with no protection other than rubber boots to prevent snake bites.

"Many of the pesticides used on cocoa plantations have been banned in Europe and the US because they are too dangerous." He points our that many cocoa workers are either illiterate and so cannot read any rudimentary instructions for pesticide use, or have not been properly trained. That produces a catalogue of problems from plantation poisonings and "accidents" to birth defects and chronic illnesses among workers.

"Advertising, together with direct pressure from manufacturers to increase pesticide usage is making the problem worse. National laws designed to control pesticide use are either too weak, or simply ignored because there are not enough people to enforce them," says Dr Hay.

The activities of transnational companies in encouraging growers to rely more on pesticides comes in for further criticism from Craig Sams, supremo at Whole Earth Foods. In 1987, he visited cocoa growers in Belize.

"The Mayan growers were still practising the traditional biological system of interspersing wild cacao plants amongst cultivated cocoa trees to strengthen their genetic resistance. Then the American Hershey Corporation came along with hybrid cocoa trees that could crop several times a year. They offered free trees and premium payments for cocoa beans if growers would rip up the wild plants and replant their plantations. Gullible farmers put the new trees in, and found that although they were higher-yielding, they were more prone to fungal disease and needed much more fertilisers.

"By 1990, the cocoa price had dropped and Hershey was not paying the same premiums. The rising costs of reliance on pesticides and fertilisers meant that farmers were actually worse off than before. Meanwhile, the sustainable agriculture system they had used for centuries was screwed up," says Mr Sams.

When it comes to the health of chocolate consumers, rather than cocoa producers or the environment, it is known that residues of pesticides do turn up in beans. Industry bodies, such as the Biscuit, Cake, Chocolate and Confectionery Alliance, argue that these are minute, and far below government safety levels. Dr Hay agrees that there is no evidence to suggest that residue levels are dangerous. But he says the question to ask is: "Do pesticides need to be there at all?"

Green and Black's organic chocolate is on sale this month in natural food stores, £1.89 for 100 gms.

Figure 5.3 Article first published in *The Independent* in September 1992. Reproduced with permission. © *The Independent*.

In 1993, Togo began to turn violent. After he lost the election, Gnassingbe Eyadema decided that, as he still controlled the army, he would not give up the Presidency. Riots ensued, the French cut off aid and the ports were blocked by unpaid strikers. Organic cocoa beans that were awaiting shipment had to be packed in air containers, airlifted out of Togo and flown to the Chocolaterie d'Aquitaine factory in Bordeaux, who contract manufactured the products and who kept a production slot open for this emergency shipment. The first production of the resulting chocolate was flown to Gatwick where a waiting van whisked the output to Sainsbury's just in time for the delivery slot and in time to avoid a delisting.

Belize

Shaken by the anxiety and expense of this experience, in May 1993 Green & Black's contacted Diego Bol, who was headmaster at the Toledo Community College in Punta Gorda town in Belize, and who came from one of the main cacao-producing villages in the foothills of the nearby Maya Mountains. He reported that things had been going downhill for the local cacao producers and he put the company in touch with Justino Peck, the chairman of the Toledo Cacao Growers Association (TCGA). Peck confirmed that the farmers were in trouble as they had borrowed money in the 1980s to purchase hybrid seeds ($1 each) and fertiliser and pesticides on the advice of agronomists from the Overseas Development Administration, USAID and the Peace Corps. This aid project mirrored similar projects throughout the developing world in the 1980s, instigated in part by a desire to find a cash crop for small farmers in tropical highlands and in part by a desire to break the monopoly on cacao held by the Ivory Coast. The cocoa market had spiked to $3000 per tonne from its usual price in the region of $1000 per tonne when the Ivory Coast, in the early 1980s, decided to withhold their crop from the market as the price was too low. Both commodity traders and the chocolate industry sought to reduce the power of a single producer country and their governments assisted via their aid agencies.

There had been bitter disputes among the producers in the villages in the Maya Mountains as reservation land that had hitherto been held in common had to be carved out and put into individual title to provide collateral to Belize Bank against loans to purchase seeds and pesticide. All this planting and borrowing had been predicated on a commitment by the Hershey Hummingbird plantation to purchase fermented dried cacao beans at a price of $BZ1.75 per lb ($US1 = $BZ2). Hershey had pulled out in 1990, selling the plantation on soft terms to the manager. The Commonwealth Development Corporation's Big Falls plantation had also decided to bulldoze their 70 acres of cacao and replant with grapefruit. The intermediary buyers from Hershey Hummingbird progressively reduced the price they were paying from $BZ1.75 to $BZ1.25, then to $BZ0.90, then to $BZ0.70 and finally to a mere $BZ0.55. At this price most farmers could not afford to spend the time to harvest and process their cacao and production dropped by more than half, just as the trees were reaching maturity. The bank was applying pressure for repayment of the outstanding loans and many farmers had to work as migrant labour, picking oranges or cutting sugar cane on plantations to the north, just to service their debt. It was a classic example of how land can be alienated from its traditional inhabitants and users and be brought onto the market. In nineteenth century USA unscrupulous merchants sold whisky on credit to Indians, then cleared the debt by taking reservation land in place of cash until Congress legislated against the alienation of reservation land. No such protection existed in Belize.

Out of this desperate situation there was an opportunity to launch a product and a project that from the outset could be the embodiment of organic and fair trade principles. The name of the product was to be Maya Gold, reflecting the prosperity that cacao had once brought to the Maya, when cacao beans were the medium of their trade and the main source of their trading wealth. It would be flavoured with a blend of spices and citrus, modified from a spicy flavour blend that had originally been developed as a flavouring for Whole Earth apple jelly and apple juice. This reflected the traditional Maya 'kukuh' drink that was drunk during the ceremonials of the Deer Dance, where cacao, enhanced with the flavours of allspice, 'tzbek' vanilla and limonoids from *Choisya ternata* tea combined to make a spiritual sacrament. A draft label design, enhanced with an image of Tlaloc, the rain god pictured on the walls of Teotihuacan (a Toltec centre but Maya-related) was presented to the Sainsbury's confectionery buyer. He was intrigued by the whole concept and agreed to stock the product on condition that Sainsbury's were the exclusive supermarket customer for a six-month period from launch. This was agreed as Sainsbury's commitment was crucial to progressing the project.

On a visit to the TCGA in Belize, accompanied by a Soil Association inspector, a new deal based on a new product concept – Maya Gold – was offered to the TCGA. The proposed contract comprised:

(1) A five-year rolling contract, paying $1.25 per pound of fermented, dried cocoa beans
(2) Help in obtaining organic certification
(3) A $US20 000 cash advance to guarantee 'spot cash' rather than receipt vouchers for the farmer members
(4) Training for key co-op members in management accounting, correct fermentation and quality control to ensure the best quality cacao
(5) An additional 5¢/lb premium for farmers who planted mahogany, cedar and cohune palm as shade trees

Before the deal was agreed and signed, British and FAO aid workers in Belize advised the Maya strongly against going organic, which they said would be a disaster. They had trained the farmers in the use of chemicals and other modern agricultural techniques and feared that going organic would lead to disease problems and falling yields. The contract offered more than double the current price, however, and despite the extra work involved in obtaining organic certification and the uncertainty that anyone from Green & Black's would ever return to Belize, the farmers decided to sign the contract. The Minister of Agriculture, Russel Garcia, had visited the co-operative and pronounced the dark chocolate 'Delicious, this is the taste of real chocolate!' and ultimately, his enthusiastic comments, combined with the offer of higher prices and a cash advance outweighed the advice of the experts.

Because the EU Regulation 2092/91 had only come in force a year earlier, the normal conversion period that applies to perennial crops such as cacao trees did not apply. As long as the Soil Association inspector, Michael Michaud, could satisfy himself that no agrichemicals had been used in the previous two years, the Soil Association could, subject to setting out criteria for compliance in the future, certify as organic the cacao from TCGA members. The inspection was inevitably unusually rigorous as it sought to establish compliance in the absence of previous record-keeping. However, because of their innate reluctance to spend money on chemicals and because of the poor prices that had prevailed, the farmers, with few exceptions, were able to satisfy the inspector. On 8 December 1993, the Soil Association Certification Committee granted certification to the TCGA, subject to a further inspection in April 1994.

On 3 December 1993, Mike Drury, Managing Director of the Fairtrade Foundation, at a press conference where Cafédirect had invited a coffee grower to talk about the benefits of fair trade, suggested that Green & Black's might qualify for the Fairtrade Mark. Simon Wright, Whole Earth's Technical Director at that time, had been involved in the creation of fair trade standards and suggested that there would be no problem in compliance. At the beginning of January 1994, Mike Drury, along with Bill Yates from Oxfam, one of the main supporters of the Fairtrade Foundation, flew to Belize to see for themselves. They returned completely satisfied. Drury's words were 'It embodies everything we were led to expect, and more.'

Maya Gold

Maya Gold was launched on 7 March 1994 on the Oxfam stand at the BBC Good Food Show in London. That date marked the birth of the Fairtrade Mark – until then it had been a worthy idea but had yet to appear on any actual products (Figure 5.4). Overnight the Fairtrade Mark became a supermarket shelf reality, starting with Sainsbury's and soon all the multiples. Sainsbury's generously agreed that, given the historic opportunity this represented for fair trade, that they would release Green & Black's from the commitment to supply their supermarkets exclusively and Maya Gold was soon on sale in Tesco, Waitrose, Safeway and other leading supermarkets, after one of the most remarkable grass roots marketing campaigns ever seen. Vicars urged their congregations to support this initiative and telephoned the Tesco confectionery buyer to exhort him to stock Maya Gold. Christian Aid newspaper splashed the story on the front page exhorting their readers to support

Figure 5.4   The Fairtrade Mark.

the initiative. Young Methodists carried a flaming torch for fair trade from town to town, running in relays, then urged store managers to stock Maya Gold. Oxfam rallied its supporters and stocked the chocolate in their high street charity shops. Hitherto those shops had dealt in donated clothing and domestic goods but had not dealt in foodstuffs.

Newsround, the BBC's children's news programme, took several cases of newly minted bars of Maya Gold to Belize and recorded the story from the producer end, filming the farmers harvesting cacao and then panning to their photogenic children tasting the unfamiliar product of their parents' production. Never before had most chocolate consumers made the connection between a bar of chocolate and a bean from a fruit from a tree grown by producer communities. This footage made it onto the mainstream BBC news and was broadcast in Welsh and then internationally syndicated via CNN. Further press and television coverage was extensive.

Within a year Cafédirect, Clipper Teas and Equal Exchange all decided that they too would commit to the Fairtrade Mark and the Fairtrade Foundation grew apace. Within eight months of the launch the Fairtrade Foundation brought Justino Peck, TCGA Chairman, to the UK to meet the press and fair trade supporters from Oxfam, Christian Aid, Cafod, World Development Movement and other organisations. Radio, television and press coverage was extensive and enthusiastic. Peck's articulate expression of the farmer's perspective, emphasising the security of income as much as the level of income, drove the message home in a way that no advertisement could ever equal, especially to the deeply committed constituency that he addressed.

Combined with the effectiveness of the organic and fair trade message, the most powerful way of converting people to Green & Black's lay in sampling. Once a person tasted the rich, dark complexity of flavour of 70% solids chocolate, they found other plain and milk chocolate insipid. Sampling at consumer shows such as the BBC Good Food Show and the Hampton Court Flower Show ensured that the target audience knew what the products tasted like so that, when they saw them on shelf, they already knew what was inside the wrapper and could purchase with more confidence.

Unforeseen benefits

The economic benefits of market security, both in price and length of contract, to the growers are obvious. A bonus has been a cascade of unforeseen additional benefits – a veritable virtuous circle.

Smallholder production of cacao strengthens the economic position of women. Because the fermentation and drying of the cacao takes place in the household area, the women take control. The cacao is sold to the TCGA in small but regular lots, so they often take it into town when they are going to market. This means that they are more likely to be the recipients of the payment for cacao. Men tend to spend money on equipment, cattle or beer. Women are more likely to spend disposable income on health and education (Figure 5.5). Previously no more than 10% of primary school children went on to secondary education. This figure has increased to 70% since 1994.

While it may seem mundane, women also use additional cash income to purchase tinned sardines. The only local and traditional source of oily fish in the Maya villages is a small freshwater minnow that is caught with corn meal placed in traps made of old bottles. Inclusion of canned sardines in the diet of preconceptual and prenatal mothers contributes significantly to healthy foetal development.

Every Maya village is sited on a river that serves as bath, laundry and drinking water supply. The reduction in pesticide run-off resulting from organic cacao cultivation, as well as the reduced dependence on pesticides in citrus and rice production thanks to an increased understanding and adoption of natural pest control practices, has led to a fall in the incidence of skin diseases, blistering and hives.

Migratory bird populations have increased dramatically, reflecting increased forest shade cover and reduced pesticide residues. The Kentucky warbler has been

Figure 5.5 Poyonaam Women's Group.

sighted more frequently, a good indicator of increased habitat availability. A scarlet macaw breeding reserve owned and operated by the Audubon Society now encourages resident farmers to join the TCGA in order to protect the habitat for macaws, which come from other parts of Central America to breed in Belize.

Maya Reservation land has been kept intact and the bank has not foreclosed on the old loans. The alienation of some bits of reservation land is probably irreversible, but the ownership is in small parcels owned by local and mostly Maya farmers.

The status of the Maya in Belize has risen – they are the only farmers in the country who are not holding out the begging bowl to government or to aid agencies and, as the TCGA expands, it brings in more and more non-Maya farmers who respect and admire the way it is run. The status of indigenous peoples in most of North and South America is low – the Maya are now a political force in Belize.

These gains are mirrored by the massive goodwill this project created for Green & Black's. The full range of products benefit from this halo effect. With a long history, this project continues to successfully embody the highest ideals of what fair trade should be about.

A consumer leaflet about Maya Gold was produced and widely distributed that set out the principles that drive and define ethical business practice (Figure 5.6).

From the outset, the core marketing technique was to give away as much chocolate as possible, in confidence that the conversion rate would justify it. Since 1991,

Maya Gold
Rainforest Treasure

* Organic
* Fairtrade Marked
* Authentic Maya taste

Figure 5.6   Maya gold leaflet.

Green & Black's have given away an estimated £2m worth of chocolate and this form of promotion continues to be the most effective way of influencing consumers to switch their brand loyalty. The security of the relationship with growers that an organic long-term contract provides means that quality control and improvement programmes can be implemented that will deliver cacao that has been harvested at the right time, fermented carefully and sun-dried. These crucial stages of post-harvest processing create the flavour precursors in cocoa beans that roasting develops and enhances. This in turn means better-tasting chocolate.

In 1994, Josephine Fairley sent bars of 70% solids chocolate to chefs and friends and asked them for recipes. The result was a 12-page recipe booklet that included Linda McCartney's brownies, The Groucho Club's Chocolate Pie, Hugh Fearnley-Whittingstall's Maya Gold Mousse, Alistair Little's Fudge Sauce and Launceston Place's Chocolate Berry Torte. Threaded through the leaflet were details about cacao production and the back cover had a picture of her and Miguel Mes, one of the Maya farmers, standing in front of his cacao trees in the village of Laguna. The accompanying text encapsulated the story so far. More than a million copies were produced and distributed, enhancing the prestige of the brand and increasing the ways it could be consumed.

Competition at last

After a clear run at what was a tiny niche in the confectionery market, it was inevitable that we would face competition. In the autumn of 1994, Lindt launched 'Excellence', a 70% solids chocolate. Almost immediately orders from Tesco dried up and within two months of the launch, Green & Black's were no longer on sale in their stores. Other supermarkets saw a brief dip in sales as consumers tried the cheaper, non-organic alternative, but then returned to their regular purchasing of Green & Black's. Lindt carved out a share of the premium, high-solids cocoa content chocolate market and enlarged the niche.

Overall it was a struggle – the total market for organic food in 2005 was £1.5bn but in 1991, the year of launch, it was a mere £92m. By 1997, yearly sales had reached £250m and at this stage Sainsbury's and Waitrose scented the opportunity and raced to become the most organic supermarket. (Safeway, in 1994, under advice from management consultants McKinsey to shed their upmarket image, had pulled out of organics almost entirely.) With the Whole Earth grocery range and Green & Black's chocolate, Whole Earth Foods were ready for them with a range of superb organic products that were well packaged, competitively priced, tested in the marketplace and were already established leading brands in the specialist sector. The supermarkets gave organic suppliers kid glove treatment at that time, conscious that they couldn't set one supplier off against another when there were so few reliable suppliers around. It was a rare moment of opportunity.

Restructuring

In 1998, however, despite invoice discounting and tight credit control, the cost of financing increasing levels of stock and debtors led to a serious cash squeeze, rapid growth was causing problems yet there was no way to slow the pace without risking loss of market share. Duerr's – the manufacturers of Whole Earth peanut butter and jam – also handled the warehousing and distribution of the products, achieving mutual efficiency and cost saving. Apprised of the situation, Duerr's offered to finance the stockholding costs, thereby reducing the cash flow pressures that were beginning to constrict the growth of the business. They could see how sales were going and could see how chocolate was going from strength to strength and calculated that they stood to gain from the ongoing growth of both brands and also from the commission on sales that was part of the financing deal. The relationship with Andre Deberdt was terminated. He had been unable to keep up with demand and was pressing for advance financing. He had formed an association with Cantalou, a large French chocolate manufacturer. This produced a potential conflict of interest as his focus shifted towards his own brands and private label business.

This move had a positive impact on gross margins as Green & Black's began to deal directly with ICAM, a family Italian company who were expert in manufacturing chocolate to a high quality standard. Contracts with Duerr's were drawn up and the new relationship was set to be signed and launched on 2 January 1999 when, a few days after Christmas 1998, another offer landed on the table.

In 1997, William Kendall and Nick Beart had sold New Covent Garden Soup to Daniels plc. They had a group of satisfied investors who were interested in backing them in another project. Beart had conducted an informal assessment of the finances of Whole Earth Foods Ltd with a view to making an offer for shares. When they became aware of the Duerr's offer of support they acted quickly and came in with an offer of cash for 75% of the shares and also a business plan that involved substantial investment in the brand, backed up by investor's loan notes that could be mobilised, if needed, to fuel growth even further. Faced with the choice of surrendering control of the business in exchange for providing the brand with the resources necessary to fulfil its potential, their offer was accepted. Along with Green & Black's, they purchased the Whole Earth brand and also the Gusto brand, a herbal energy drink that was marketed by Rima and Karim Sams. Duerr's were notified of the change of plan and the distribution and stock finance contract was never signed.

The new team, with several million pounds in reserve to finance the costs of sales, marketing and expansion of stock, set about making their mark on the brand (Figure 5.7).

One of the first changes was that the Josephine Fairley signature came off the packaging, so that it would escape the burdens of a personality brand and be easier to sell in the future. The investors backing them had invested under an Enterprise Investment Scheme and were looking for an exit, either a trade sale, management buyout or flotation, after no less than five years.

Figure 5.7 Green & Black's range, showing rebranding.

They created the role of President – moving Sams to a non-executive role. Despite disclosure and despite the time spent studying the business, there was still a lot of information that wasn't recorded and that was essential to an orderly transition to the new ownership and management.

The brand was traded for everything it was worth: bogofs (buy one get one free), coupons, price-marked packs and promotional offers. Neil Turpin, the Sales Director, formerly Key Account Manager for Lucozade at Glaxo Smith Kline, began to worry that he could only trade the brand so hard and needed marketing support.

The Marketing Director was Caroline Jeremy, a former recipe development expert and colleague of Kendall's from the New Covent Garden Soup Company – she redesigned the Whole Earth packaging, reformulated and repackaged the Gusto herbal energy drink and developed new kids' products for the Green & Black's range. In some ways it was more difficult for her team to focus on three different brands; previously Josephine had driven the brand management of Green & Black's, Rima and Karim Sams did the same with Gusto, and Whole Earth was Craig Sams' baby. Inevitably, not all the new initiatives worked and there were also problems with coordinating sales and marketing. It is easy to underestimate the skills that one is seeking to replace in taking over a complex operation, especially an entrepreneurial business where the structure and guidelines for decision-making had been intuitive and fluid where they existed at all. Caroline arranged 'brand essence' gatherings where the mission statement and strategy for each of the brands was set out, after consultation with focus groups comprising typical target customer groups and marketing experts. This varied from the previous approach, which had always been based on extrapolating from personal tastes for which no product existed, confident that once an appropriate product was on offer the marketplace would readily adopt it. This works for innovative entrepreneurs, but couldn't work in a more formal environment.

Caroline took maternity leave and Mark Palmer, formerly head of marketing at Burger King, was recruited as Marketing Director. He worked hand in glove with sales to make sure that product launches were delivered on time and that advertising supported trade activity seamlessly. Caroline did not return to the job, but later concentrated on producing 'Unwrapped' a high quality book that told the story of the company's relationships with suppliers and the underlying quality message of Green & Black's, all threaded through pages of sumptuous recipes using Green & Black's chocolate. It expanded on the strengths of the celebrity recipe leaflets and on the thousands of recipes that had been provided as competition entries and just out of sheer enthusiasm by loyal customers. With luscious photography by Francesca Yorker, the book enjoyed high sales from its launch and the publisher regularly produces new editions, including an Americanised version.

Crucially, the independent natural foods retailer was never forgotten. Despite a share of less than 1% of the total grocery market, they represented a good 25% of organic chocolate sales. When they could offer Green & Black's chocolate at 99p instead of the usual price of £1.55, albeit for a limited period, they could stock up and thumb their noses at the supermarkets' everyday lower price of £1.35. Their eye level till-side merchandising and reluctance to stock the offerings of competing later entrants more than rewarded our support.

Starting with a £100 000 tube card campaign, money started to pour into above the line marketing. The initial campaign was successful and repeated three more times: two cards on every train in the entire London Underground system, plus escalator posters where there was a stockist near the station.

The emphasis was on the darkness and high cocoa solids content of the chocolate, even the milk chocolate.

Sponsorship of English Heritage concerts in conjunction with Classic FM ensured that the brand reached its target audience on summer evenings, where visitors were offered chocolate and ice cream, so that regardless of the weather, they were enjoying something.

Wine connoisseurs were targeted. Their vocabulary for describing fine wines is almost identical to the descriptors used to describe the taste of chocolate. A co-promotion with Cockburn's Port, where chocolate and port were sampled in store in the Thresher's chain of wine merchants was a huge success and led to continuing listings in off-licence stores. It reinforced Green & Black's claim to be the connoisseur's chocolate.

Supply chain management

Prior to 1999, the relationship with Andre Deberdt led to increased costs, erratic supply and a tendency to adapt to what suited the supplier rather than a customer-driven approach. With the direct relationship with ICAM things began to improve. John Kennedy, a graduate of the INSEAD business school in Fontainebleau, where William Kendall was a part-time lecturer, applied directly to Kendall to work with the company and was appointed to the position of Commercial Director. He focused on improving quality, reducing costs and raising service levels. The easy routes to saving money such as squeezing producers and diluting quality were eschewed, the key lay in maximising the performance of the many third parties on which the successful delivery of the product relied. ICAM were targeted and the sometimes fusty ways of a second-generation Italian family business were analysed and assessed with the keen eye of an MBA. Kennedy's regular visits to the factory turned up ways of achieving efficiencies, reducing waste and optimising the use of ICAM's assets and workforce. The understanding of their costs, combined with the detachment of not being an owner or manager of the business, enabled Kennedy to recommend improvements and to garner the savings and incorporate them into the transfer cost of the products. Transport, warehousing and distribution all benefited from this analytical approach and the steady reduction of costs, combined with improved service levels and a high level of quality control, underpinned the efforts of the sales and marketing teams. If a batch of chocolate failed quality control because of a flavour taint or some other shortfall there was no conflict of interest such as is experienced by a manufacturer who can be tempted to overlook a product defect in the interest of saving on write-off costs. With high cocoa solids chocolate the slightest defect can be acutely noticeable – if a batch of chocolate failed to make the grade, ICAM would take the hit. This led to a careful and professional approach to quality assurance. Once a cost had been taken out of the system, if it crept back in the totality of that cost was borne by ICAM as the transfer price had already been reduced accordingly. This detachment and overview of the entire process represents one of the great advantages of outsourcing – the contract processor benefits from it as the amelioration of their inefficiencies benefits all their other production as well. John Kennedy's contribution to the success of the business is inestimable, the backroom team is a vital part of the mix and often faces the most demanding challenges.

Evolution

In 2001 an unexpected offer to purchase the Whole Earth brand came from Gene Kahn, who had recently sold Small Planet Foods to General Mills and wanted to use the Whole Earth brand as the vehicle for their European expansion plans. This put the Whole Earth brand into play. A private investment bank that specialised in food industry mergers and acquisitions was appointed to help assess the options. They soon found that there was considerable interest in acquiring the Whole Earth brand and in 2002 it was sold to Kallo Foods, the UK subsidiary of Wessanen, the Dutch multinational that had recently sold $US1bn of its US dairy industry assets and had earmarked the cash for expansion in the natural and organic foods sector.

A by-product of these wider sale discussions was an approach to Cadbury Trebor Bassett, the UK confectionery manufacturing subsidiary of Cadbury Schweppes. After considering the portfolio of brands, they decided there were insufficient synergies to justify interest in the Whole Earth brand or in the Green & Black's brand. Cadbury Schweppes, however, with a wider worldview, decided to look more closely and this led to their taking a 5% share in Green & Black's. They had complex soft drink bottling and licensing arrangements that would have made an investment in the overall business unlikely as the Whole Earth soft drinks range and the Gusto range would have had to be dropped before Cadbury Schweppes or any of its subsidiaries could have invested. The agreement included an amount for the purchase of the shares and a loan on favourable terms of three times the share purchase price to enable the company to further fuel the growth of the brand. This cash, combined with the income from the sale of the Whole Earth brand to Wessanen, provided an increased marketing budget, delivering advertising and promotional support for new product launches and for consolidating the position of flagship products such as the 70% solids dark chocolate and Maya Gold. A further aspect of the agreement was that each side had put and call options that could be exercised in a few years' time to bring about the sale of the remaining 95% of the shares to Cadbury Schweppes. The President, the CEO and the Finance Director all had holdings in excess of 10% and smaller holdings motivated the Sales Director, the Commercial Director and the Marketing Director. A capped staff incentive scheme was put into place in March 2004 that encouraged dedicated commitment from every member of the organisation. By December 2004, the target sales and margin figures that triggered the optimal purchase formula had been met and by March 2005 it was clear that the ideal moment had come to exercise the put option. Easter came late in 2004 and early in 2005, which contributed to particularly high sales figures. Cadbury's notified the company that they intended to exercise their call option and the accountants went to work to confirm the final price. On 11 May 2005, Green & Black's Ltd became a directly and wholly owned subsidiary of Cadbury Schweppes. Craig Sams agreed to remain in the role of President.

Press coverage of the acquisition was extensive, going far beyond the financial pages. Concerns were raised that Cadbury's would change the fundamentals of the brand in respect of organic ingredients and fair trade and there was even speculation that they would introduce their recipes to the brand offering. The public relations challenge was met with press interviews with William Kendall and Craig Sams and an advertising campaign that emphasised even more strongly the fair and organic credentials of the brand. Cadbury's were deeply aware that the eyes of the world were upon them and were keen to give Green & Black's as much independence as possible. A new financial director was the only immediate appointment and a new

CEO was identified and took over in January 2006. The choice was Ward Crawford, former CEO of Cadbury's Japan, an executive with a reputation within the company for thinking outside the box and also a dedicated 'green', with a 5-acre smallholding in Devon run along organic lines.

Growth

By 2006, Green & Black's had an 8% share of the UK market for bar chocolate and was still growing at 70% per annum. In Waitrose in 2006, Green & Black's represented more than 30% of their block chocolate sales, outperforming both Waitrose own label and Cadbury's. The US market in 2006 had begun to respond to the messages of both organic and fair trade. By mid 2005, Green & Black's had become the US market leaders in both the organic and the 70% solids categories, according to SPINS data. The brand has gone beyond the natural foods supermarket chains and is distributed via upscale mainstream supermarkets and enjoys distribution via Target and Wal-Mart, the leading discounters.

Quality and flavour – smallholders versus plantations

The flavour and quality of chocolate depends upon the quality of the raw material, the cocoa beans, from which it is manufactured. Producing fine flavour cocoa depends on three key stages: harvest, fermentation and drying. At all three of these stages smallholder producers achieve superior results compared to plantation operators. This is directly related to the fact that they are independent smallholders using organic farming methods. The success of Green & Black's derives primarily from its ability to consistently offer superior tasting chocolate, regularly winning in blind tastings. One does not have to be a connoisseur to appreciate the difference – vinegary, astringent and excessively bitter tastes are repellent to most people. In a highly concentrated chocolate their impact is exaggerated. As cocoa butter counts as a 'cocoa solid' in calculating the percentage of cocoa solids, a 70% solids chocolate can be made with 55% cocoa mass and 15% cocoa butter, thereby diluting the off tastes and allowing for a palatable product that claims 70% cocoa solids. Green & Black's have always maximised the cocoa mass content, only adding cocoa butter at a level of up to 3% in order to get the necessary 'snap' or texture.

The cocoa tree (Figure 5.8) produces fruit pods, each of which contains about 30 cocoa beans surrounded by a sweet, milky pulp. In natural conditions the seeds germinate in the pod and, when the first leaf cotyledons and rootlets are formed, the entire pod falls to the ground, rolls away from the parent tree, and all the seeds grow up together to form one new tree. It is a survival mechanism developed and honed in the harsh conditions of the forest understorey. As the seeds germinate, enzymes are released, which transform the stored proteins of the seeds into nutrients for the

Figure 5.8   The cacao tree.

sprouting plant. These proteins are reduced to peptides, which provide the flavour precursors that are developed in roasting. This enzymatic process, inaccurately called 'fermentation', is crucial to the flavour of the finished chocolate.

A farmer will pick the pods as close as possible to the point where they are ready to germinate. The beans are then fermented in small batches for 5 days and are then sun-dried for another 5–7 days. When this process is completed the previously purple beans have become brown inside. The industry standard for well-fermented cocoa beans is 85%, measured by sampling 20 beans from a sack and finding that more than 17 are not purple. A good quality smallholder cocoa can test at 95–97%, with some farmers achieving 20 out of 20.

On a plantation, trees are planted more closely together as management is by an overseer with a team of workers, managed by vocal commands and line of sight. The trees are planted 8 feet apart, compared to a minimum 15 feet on an organic holding. There is minimal shade, the upper part of the trees provide shade for the lower part, disease spreads and dependence on chemical fertilisers and fungicides is universal. The productivity of a team of workers is measured by how many pods they can pick in a day, so they pick pods that are underripe, often leaving them to sit for weeks waiting for them to reach the stage where some of the beans might germinate. An organic smallholder can multitask; if the pods aren't ready to harvest, he can weed beans, plant corn, tend bees, harvest squash or just take a well-earned rest. Thus only pods that are ripe get harvested.

The large bulk fermenters of plantations yield a monotony of flavour compared to the individual differences that generate complexity of flavour when hundreds of smallholders each vary slightly the way they ferment their beans. Plantations use heat-assisted drying. This shortens the drying time and beans can be down to the requisite 7% moisture level in a couple of days. Time is money, but it is also flavour. The enzymes don't have enough time to complete their work.

The result is a flavour gap between industrial scale growing, harvesting and processing of cocoa compared to organic smallholder production. This difference is most manifest in a high cocoa solids chocolate and the success of Green & Black's derives from the fact that it has established enduring and mutually beneficial relationships with small farmers worldwide who are committed to providing the best quality cacao to justify the price premium they are paid and because they take pride in the finished product. They regularly receive samples of the chocolate made from their beans, a rare experience for the majority of world farmers.

The social, political and economic benefits of a society of landowning small farmers with reliable incomes are immeasurable. Political stability is undermined when workers are kept in poor conditions and paid minimal wages. In the case of cocoa production, the industrial model barely works and many large plantations have been unable to compete with small producers.

The successful production of delicious chocolate depends on success in farming organically, in processing after harvest with care, in establishing fair relations to ensure loyalty in the supply chain. Respect for the culture of the producers is repaid with quality. It is serendipitous that today's consumer expects a high standard of corporate behaviour and that adherence to this standard fulfils the consumer's expectation of a high standard of taste. This happy interdependence has ensured Green & Black's success in the marketplace.

Chapter 6

Case History: Abel & Cole

Ella Heeks
Managing Director, Abel & Cole

Introduction

Abel & Cole is an organic home delivery retailer. Since 1993, the company has provided boxes of seasonal fruit and vegetables. Customers order a suitable size of box for their families; the contents change each week to reflect all that the seasons have to offer. Abel & Cole also offers a broad range of organic products, including meat, fish, dairy, bakery, juices, wine and beer. Abel & Cole does not have any shops; it only sells goods for home delivery. Orders come from customers by phone, email and through the company's website. Customers can order on a one-off basis, but many prefer to place rolling orders, which means goods are sent out regularly – usually every week.

Abel & Cole has enjoyed exceptional success, averaging over 70% growth every year for the last five years. This significantly outstrips the average growth of the organic retail market (12% annual average)[1] and even of the thriving independent sector (18% annual average)[1] during the same period, as shown in Figure 6.1.

As the organic market has grown, an increasingly broad range of consumers have joined, bringing with them a set of expectations and requirements formed by shopping habits and lifestyles, which are more mainstream than the initial organic consumers. This development in the marketplace is paralleled and illustrated by that of Abel & Cole's own customer base, discussed in the next section. Abel & Cole's success in meeting the demands of the broadening customer group can be explained by four key factors:

- Products that meet the needs of organic consumers
- Sophisticated service
- Procurement policies that reflect consumers' values
- Ethical practices that extend through all areas of the business

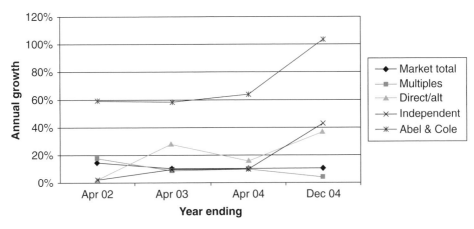

Figure 6.1 Comparison of annual growth rates and organic retail sales annual percentage growth.

Some of these factors were in place within the business before the market started to grow; others have been developed in response to the new needs and expectations that have been presented by the broadening market.

Customers

Abel & Cole started delivering organic food to Londoners' doorsteps in 1993. At this time, organic food was little known and certainly not available in the supermarkets, and the total market was worth around £100m per year. Demand for Abel & Cole's services was from 'deep greens' – people whose lifestyle was defined by their concern for the environment, and who would dedicate significant amounts of time and care to recycling, to reducing household waste and frequently to campaigning on environmental issues. Abel & Cole's customers were defined by their values much more than by income, age or gender – the company was equally popular in the poor neighbourhoods around its Brixton warehouse as it was in the wealthier areas of London.

Abel & Cole still has many 'deep green' customers, who are attracted to the company by its strong environmental credentials. Since 2000, though, Abel & Cole has succeeded in attracting a much larger and broader group of customers – a direct reflection of the broadening appeal of organic food. To some extent, this is explained by the fact that the environmental concerns that motivated Abel & Cole's first customers are now shared by a much larger group of people. In addition, a broader set of motivations are now bringing people to Abel & Cole, including health concerns, the trend for seasonal cooking, the convenience of the service, anticorporate sentiment, a response to a friendly brand and the perceived cachet of buying one's provisions from Abel & Cole. 'Deep greens' now form a minority of Abel & Cole customers. The majority have joined the service for a mixture of reasons, including environmental reasons, but their lifestyles and interests do not revolve around the environment.

Communication

As the Abel & Cole customer profile has changed, so have the expectations that the company has needed to meet. Abel & Cole has been successful at keeping in touch with its customers and understanding their needs. The company maintains its awareness of its customers through effective two-way communication. Abel & Cole makes the first move, sending out a newsletter to every customer every week, which explains where the produce they are buying comes from and gives information about the environmental and ethical initiatives that the business is taking. A sample newsletter is shown in Figure 6.2.

Crucially, the tone of the newsletter is simple, friendly and respectful; it does not read like a corporate newsletter. Another rule is that Abel & Cole should not take centre stage in the newsletter; the aim is to serve as a conduit between the grower and the customer, and to give the customer an opportunity to learn about the origins of the food they are eating. A typical article is shown below:

The Green Fingered Man

This week we've been learning about the Green Man of Sherwood Forest, otherwise known as Jack in the Green, Green George and the May King. He's a symbol of the spirit of nature, and as the May King he's the symbol of the summer's battle against the evil winter. Now, we don't believe that the winter is evil, but we do think that the May King has sneaked on to Haywood Oaks Farm on the edge of Sherwood Forest in Nottinghamshire because Charlie Anstey seems to know exactly what to do to stave off the wintery frosts that might damage his carrot crops.

Carrots burrow down into the soil like any other self-respecting root, but their tops, or crowns, always poke above the ground. If a frost descends on the fields then this can do the crowns some serious damage, and will make them go all mushy – maybe winter isn't that nice after all! It doesn't matter to Charlie though, because Charlie knows some clever little secrets. Around November he puts down a layer of straw over the top of his carrots, which makes a kind of duvet for the little orange roots. This straw bedding stops the carrot crowns from getting icy, and also stops the ground from freezing, which makes it a lot easier for Charlie to harvest his vegetables.

This isn't the only way that winter seems to magically disappear on Charlie's farm. Most farmers have to watch the winter weather very carefully before they get their wellies on to start their harvests. Heavy rain at this time of the year can easily clog up the soil and make it too muddy for any machines to help with the harvesting; it doesn't make it that easy to do by hand, either. Fortunately

ABEL & COLE

Week commencing 25th September 2006

Man versus badger

If you thought organic farming was tough (as well as rewarding, of course!) how about this for a challenge? Plant a sweetcorn crop in May, nurture it without artificial fertilisers, man-made chemicals or pesticides in the unpredictable English climate, pray that the crop isn't flattened by hail, scorched by drought or frozen solid by frosts, and then finally reap the fruits of your labours at the end of September after five months of nail-biting toil.

Oh, but that's not even the half of it. Mike Smales, of Lyburn Farm in Wiltshire, who grows some of this week's 'Mainstay' sweetcorn, finds his organic crop under attack at every opportunity. Not only from sparrows "who leave a terrible mess", but from deer, dogs, badgers and, worse still, light-fingered humans, who pillage, plunder and scoff his beloved crop with carefree abandon!

He must be tearing his hair out. "We have a lot of damage. The dogs pull off the leaves. Deers chomp all of it, leaves and all," says Mike. But they don't particularly faze him. The real pests are the badgers who trundle down in their hundreds to the sweetcorn patch at night for a bit of a corn-fest.

"Badgers love sweetcorn – they get a cob down, pull it off and nibble the kernels just like we do! You don't want to take a badger on – they can be very stroppy, just as vicious as a big dog if cornered." Right then!

Even more frustrating than the corn-gobbling badgers, however, are the humans. "People stop at the roadside and fill up carrier bags of corn," says Mike. "They fill one bag up, put it down, then move to another row and get lost. The rows all look the same so they can't find their way back to their first bagful!" Can't say we're *that* sympathetic...

So savour each and every mouthful of your sweetcorn this week – Mike chased off a whole menagerie of ne'er-do-wells to bring us what he describes as "a big bold cob", and it's the most delicious and juicy corn in England. Doesn't his victory taste sweet indeed...

Venison is here!

Our mouthwatering venison sausages and burgers from Daylesford Organic are back on sale in time for the British game season. Succulent and full of flavour, venison is a lean meat choc-full of iron and omega 3 fatty acids. Go on, treat yourself! From £3.92.

Celebrate the best of British

It's got it all: celebrity chefs cooking up gourmet delights, food tastings and loads of other foodie events. But what is it? British Food Fortnight, 23 September to 8 October, at venues nationwide. For info, see www.britishfoodfortnight.co.uk

Bite-size nutrition

Simply bursting with crunchy goodness, beansprouts keep on growing after harvesting, so they actually increase in nutritional value! They are rich in vitamin B, iron and folic acid.

Fresh tips

Our fresh herbs will last up to a week in the fridge. Basil, chervil, chives, coriander, dill, mint and parsley all freeze well. Chop, spread out on a tray to freeze, then store in bags for up to six months. Oregano, rosemary, sage, tarragon and thyme are best dried hanging upside-down in a paper bag then stored in air-tight containers.

Variety is the spice of life

The William pear's stout form makes it the archetype of all pears. When ripe it is yellow in colour. The flesh is succulent and sweet, which makes it great for cooking.

Soil Association lecture

If you read 'Fast Food Nation' and were horrified by its exposé of the fast food industry, you might be interested to know that its author, Eric Schlosser, is giving a talk entitled 'Food Quality, Human Health and the Threat of Industrial Agriculture' at the Soil Association's 60th anniversary Lady Eve Balfour Memorial Lecture. Date: Wednesday 25 October. Venue: Central Hall, Westminster, 6pm. For info, see www.soilassociation.org/ladyevelecture

Phone: 08452 62 62 62 Web: www.abel-cole.co.uk Email:organics@abel-cole.co.uk

Printed on chlorine free, recycled paper

Figure 6.2   Abel & Cole's weekly newsletter.

for Charlie, Sherwood Forest sprung up on very sandy soil as it used to be covered with rivers that dropped sandstone deposits all over the place, and unlike clay soil, sandy soil doesn't hold water very well. This means that when it rains, the water drains away very quickly and Charlie can harvest whenever he wants! Definitely sounds like the May King is helping him out to us!

The final clue that makes us suspect Charlie knows the Green Man of Sherwood is all the grass strips, hedges and trees that he's been planting to help the insects and birds thrive – the forest is even creeping back into his fields![2]

The tone of the Abel & Cole newsletter is followed by all of the company's other communications, particularly the website, emails and the phone. While the newsletter and the website can be created by a trained marketing person, a skilled and well-managed customer service team has been crucial in extending the tone through to emails and the telephones. Carrying it through in this way is utterly essential, because it is by email and phone that the customers respond. Abel & Cole's understanding of its customers doesn't come from big surveys or focus groups; it comes from hundreds of conversations with customers, which happen every day. The conversation is always started by the customer, but it is an indirect response to the communications that have come from Abel & Cole, all of which send the message that the company is at the service of its customers and is receptive to their input. Abel & Cole's customer service team are empowered and motivated to listen carefully and to pass the customers' ideas, requests and complaints into the business so that it can respond. (More detail on how this is achieved is given in 'The customer relationship'.)

The next step is to ensure that customer feedback is met with a quick and effective response. If the customer sees that Abel & Cole has responded, he or she will be encouraged to give more feedback in future. In the early days, when the founders of the business were answering the phone, or at least relaxing after work with the people who answered the phone, the feedback loop was very simple. As the business has grown, IT systems have been set up to ensure that every piece of customer feedback is instantly available to the department that needs to respond, and management processes have been put in place to ensure that response happens.

Anticipation

Gathering and responding to customer feedback, however, only explains part of Abel & Cole's development. The other key to success has been an ability to anticipate what customers will want, well ahead of time. For example, sustainable fishing started to receive significant attention in the press in early 2006; Abel & Cole introduced its first Marine Stewardship Council certified fish in early 2003. Similarly, biodiesel became a buzzword in mid-2005; Abel & Cole had introduced it in early 2004. This ability to anticipate the issues has meant that by the time customers raise the issue, Abel & Cole is already abreast of it, and is thus able to reassure the customer and reinforce their confidence that Abel & Cole shares their values. There is one extremely simple reason for this: Abel & Cole does share their values.

People who work at Abel & Cole are motivated to do so by the same factors that motivate their customers to shop there. This makes the process of product and service development incredibly simple. Throughout the operation, there are people who care passionately about good food, reducing waste, looking after the environment and providing good livelihoods for British farmers. The people who like good food tend to gravitate towards product development; those who are passionate about farming end up in the buying department. In each area, therefore, the business is being shaped by people who want the same things as the customer; because they are inside the business, they will tend to spot opportunities and make the changes they want to well before customers start to ask for them.

Abel & Cole tells its customers about the ethical developments it makes in its product range and business practices. This may well influence customers' expectations of other food retailers, and thus contribute to good practice in the marketplace.

Range

Box schemes

In 1993, when Abel & Cole started selling organic food, the range of products available in the UK was very limited. Abel & Cole sold the broadest range that it could, which in practice meant boxes of vegetables. Box schemes, where a customer signs up for a weekly delivery of a certain size and value, but where the contents of that package changes each week, were virtually the only way that organic food was retailed in the UK until the mid-1990s. They have been used by farms for decades as a simple way to sell produce direct to local customers with minimal administration or waste. Abel & Cole's use of the box scheme format was born of necessity: it simply wasn't possible to offer the same items from week to week, because supplies were so limited and erratic. Abel & Cole's current best-selling box is a mixed selection of fruit and vegetables, which is pictured in Figure 6.3.

While born of necessity, the box scheme format had many merits:

- It provided a perfect vehicle for seasonal produce. Awareness of produce seasons has declined a great deal because of the year-round fresh produce offerings in the supermarkets. By purchasing a box, a customer could leave the experts to select seasonal produce, rather than spend time trying to figure it out for themselves in a supermarket aisle.
- It provided variety – taking people away from their habitual purchases, encouraging them to try new things
- It involved minimal, reusable packaging

Figure 6.3   Abel & Cole's best selling mixed box.

● It was well suited to repeat or regular purchasing. From the outset, Abel & Cole sold its boxes as weekly or fortnightly offerings that would reappear on the customer's doorstep at the agreed frequency unless the order was cancelled. This had great benefits for the customer: just like milk, in most households fruit and vegetables are consumed every day and are perishable and difficult to transport. Taking a regular delivery makes sense for all these reasons. For the business, the regularity of the box scheme system has filtered through into a staggeringly high level of repeat business; 95% of people buying from Abel & Cole in any given week will also buy the following week.

Whilst box schemes have their origins in the early 'deep green' market, the benefits outlined above also met the needs of many new organic consumers. This helps to explain why box scheme sales still represented around 10% of the organic retail market in 2005.[1] Abel & Cole has found that people who are used to supermarket shopping, however, do need several weeks to get used to box scheme shopping. They need time to get used to planning their cooking in response to their shopping, rather than vice versa. Once that time has passed, customer feedback such as the following is typical:

'The variety of fruit and veg sent means we have learnt some new recipes and been more adventurous!'(Abel & Cole customer email, 31 January 2006);

'...it is fun to try things I wouldn't buy easily myself. The recipe page on your site came in handy!'(Abel & Cole customer email, 15 March 2006).

Abel & Cole box service

Abel & Cole has kept the box scheme format at the heart of its service, recognising that it works well for the customer and the company. Over the years, however, the format has developed in anticipation of, and response to, customer needs. The evolution of the Abel & Cole box is a good illustration of how the business develops.

(1) **Exclusions:** after a couple of years receiving deliveries of his own selection boxes, Keith Abel started to feel that it was rather unreasonable to insist that his customers take his weekly selection whether they happened to like everything in it or not. He decided that Abel & Cole would only send out produce that each customer would enjoy eating. This was easier said than done, of course, because each person has different tastes. Customising each customer's box manually would be impossible for more than a handful of orders each day. Keith hired an IT programmer and set about making a system which would allow him to keep track of customers' preferences, check the weekly selection against it, and customise boxes for those customers who didn't like something planned for that week, replacing it with something they did like. The IT system was essential; as the scale of production grew, it would have been impossible for Abel & Cole to manage exclusions manually; it would have been forced to restrict its growth or to withdraw the exclusions service.

(2) **Exclusion groups:** several years later, Abel & Cole improved its exclusions service to make it easier for customers to use. Now customers could, at the touch of a button on the website, exclude little gem lettuces or all lettuces, Conference pears or all pears. It was a small change in comparison to the initial construction of the exclusions service, but it is a good illustration of how closely Abel & Cole was listening to its customers and how finely the service was tailored in response.

(3) **Varying frequency:** the perishable nature of the fresh produce in the box means that most people will find a weekly delivery helpful. In the early days, the majority of customers had quite stable lifestyles, spending their evenings at home, cooking most nights, often for their children. As the market broadened, more young people started to buy from Abel & Cole. They tended to have less predictable lifestyles, travelled more for their work, and socialised more. In response, Abel & Cole offered customers the option to take a delivery every two or three weeks, or even less often. This enabled them to ensure they got some good food into their households, without tying them to a routine of cooking every night.

(4) **Box range:** Abel & Cole offers a choice of 15 different types of box; most competitors offer three to five. The range has been broadened in response to the demands of customers over the years. In particular, Abel & Cole has added many selections to cater for smaller households and for more adventurous cooks (a deluxe box, also available in a small size; a salad selection; juicing bags). These changes reflect trends in society more broadly and also highlight the broadening of the organic market. The broadening of consumer demand has coincided with a broadening of the produce available from British growers, which has made it possible for Abel & Cole to offer this wider range.

(5) **Custom boxes:** the variety that appeals to some customers is tiring for others. Conversations with customers who were cancelling their box orders told Abel & Cole that some people just didn't like having somebody else choose their fruit and vegetables for them; they wanted to choose for themselves. Generally, these customers wanted the same things every week – perhaps they had a large family whose various preferences left only a few safe items; perhaps they valued routine; perhaps they just really loved apples. In response to this feedback, in 2002, Abel & Cole developed a pick-your-own service, where customers could put their own selection of fruit and veg together. Whereas this wouldn't have been possible in the early days, the range of organic produce available week-in, week-out was much broader by 2002. Nonetheless, Abel & Cole does not air freight, so the company had to find a way to deal with regular orders when the season ended. Dealing with end-of-season and blips in demand is a particular challenge for a company dealing with regular orders; food shops can just omit out-of-season items from the shelves.

Together, these enhancements help Abel & Cole to attract and keep customers. The most important attraction of the Abel & Cole box, though, is the promise that it is not just organic but ethical in all respects: the produce will be bought at a fair price from the grower, will never be air freighted, and will be packed and delivered ethically too. This is discussed further in later sections.

Broader range development

Abel & Cole started developing its range in earnest in 2000 in response to growing numbers of enquiries from customers regarding the availability of organic meat in particular. Range development is one area where Abel & Cole has used more systematic methods of collecting customer feedback. Tastes vary so much that it has proven unwise to go with the impulses of staff – a lesson learnt early on when a consignment of artisan organic chutney that Keith Abel had fallen in love with on a farm visit failed to appeal to customers. It also made a rather odd entry in the catalogue alongside all the fruit and vegetable boxes. Similarly, many requests are made by customers who are avid fans of particular products that are hard to get elsewhere, but that may not get much uptake from the customer group as a whole.

The priorities for range development have therefore been set using customer surveys. General surveys are only taken every couple of years; each one generates enough information to provide two years' worth of product development. Perhaps because it requests feedback so infrequently, Abel & Cole has found the response to its surveys to be exceptionally strong, at around 70%. This enables the company to set the priorities for range development confidently on the basis of the survey results. When a particular range is being developed and selected, the customers who requested it in the general survey are invited to give more detailed information about what they would like.

Abel & Cole has kept moving with range development, increasing the number of lines offered from 60 in 1999 to over 300 in 2006. It has particularly broad offerings of organic meat (64 lines) and fish (32 lines). Nonetheless, its range remains far narrower than that of a supermarket or large independent retailer, and there are entire categories (e.g. prepared food) yet to be developed. The relatively slow pace of development is primarily the result of the very rigorous selection process that the company uses for each addition to the range. The first principle is that there should only be one of each item rather than several competing brands: the company should weigh up all the criteria and offer their customers the best option. The second principle is that the product should always be bought as directly as possible – any party standing between the company and the producer must be adding clear value to the process, not just facilitating a deal. Beyond this, each possible product must be evaluated against a fixed set of criteria, researching this is very time consuming. The company's current criteria cover producer certification, ethical credentials, supply chain, food miles, transportation, product quality, environmental packaging, financial considerations, marketing and other factors.

Service

Abel & Cole's strength is based on its service as much as its products. Abel & Cole's approach to service starts with a wish to leave the customer feeling happy after any contact they have with the company. This is put into action through the kind of relationship Abel & Cole builds with its customers, and through the practical things Abel & Cole does to make things as convenient as possible for them.

The customer relationship

On a typical day, Abel & Cole will receive several emails from customers who are writing simply to say how happy they are with the company. Many of these focus on the quality of care they get from the Abel & Cole customer service team. For instance:

'I have been consistently impressed with your telephone staff. Any alterations to orders, complaints or requests have been dealt with in a friendly, professional and efficient manner. I am not always easy to please, but am extremely impressed with the service I have received so far. Your staff are polite and will be the reason why so many customers remain loyal to your company.'(Abel & Cole customer email, 2 April 2006)

'I would like to say that after recent events in my retail life you have restored my faith in GOOD customer service – I shall therefore continue to shop with you, happy in the knowledge that if a mistake is made someone owns up to it and does something immediately to put it right – Congratulations!!'(Abel & Cole customer email, 7 February 2006)

Unfortunately it is not possible to check anecdotal pieces of feedback such as these against a broad survey of customer opinion, as the company feels that customer satisfaction surveys (as opposed to product range surveys) would be at odds with the personal relationships it aims to build. Nonetheless, the overwhelmingly positive feedback the customer service team receives contrasts strongly with the feedback given to many other businesses, and suggests that Abel & Cole is exceptionally strong in this area.

The foundation of Abel & Cole's good relationship with its customers probably lies in the fact that the directors of the business are extremely interested in customer mood and feedback. Customer complaints are monitored daily and circulated to all directors, managers and the 23 customer service team members. The customer service manager was selected on the basis of genuine customer focus and excellent management skills. Through her recruitment decisions, she has created around her a team of people who are just as passionate about the satisfaction of the customer as the directors of the business are, if not more so. They also tend to be committed to the values of the business, and they therefore have a natural affinity with the customers and respect for their concerns about their food and where it comes from.

Recruiting people who genuinely care about the customers and who have the skills to create a good rapport with them is only part of the equation. It has been equally important to give those people the freedom to make their own decisions about how they help and communicate with the customers. For example, there are no scripts for phone calls and there are no rules about how complaints should be handled. There are also no requirements to sell more to the customers. The team are trusted to use their judgement in all cases. This, in the company's belief, is absolutely essential if they are to create genuine relationships with the customers, and to maintain their enthusiasm for the work that they are doing.

A third part of the equation is to provide the customer service team with adequate training and support to ensure that they are in a position to make good decisions and treat the customers well. The Abel & Cole customer service team has developed its own customer service training programme, which is updated and delivered by team members themselves.

Web trading and the customer relationship

Five years ago, Abel & Cole's contact with its customers was almost completely by phone calls. Since the launch of a fully interactive website in July 2003 the nature of the contact has transformed: now only 15% of transactions are conducted over the phone, with another 15% by email and 70% by the web. Customers who set up their first order with Abel & Cole through the website will also tend to manage subsequent orders that way, and around 20% of Abel & Cole customers have never emailed or spoken to the customer services team. This deprives the team of the opportunity to show these customers that they are valued. It is harder to achieve the goal of leaving customers feeling happy when they are looking at a screen than when they are talking with a positive, well-motivated person on the phone.

Abel & Cole identified this potential pitfall before it started building its e-commerce site. It was felt that development of the site must proceed nonetheless, as it would offer such convenience to so many customers. All text on the site was written with the aim of replicating the friendly, honest and respectful tone present in the conversations of the customer service team. The difficulty is that this sort of tone tends to be quite chatty and wordy, whilst web users don't like to read too much text. Abel & Cole still grapples with this challenge, and parts of the website remain rather wordy as a consequence.

Abel & Cole's second line of defence against losing a personal relationship with its web customers has been its delivery drivers. A customer who orders via the web may never speak to one of the customer service team, but may well see their delivery driver every week. The ability of the drivers to build a good relationship with customers and represent the company well has become more important than ever. In response to this, drivers are being given more training and support from the customer service team. Abel & Cole selects drivers who will enjoy the customer service side of their work, and some go on to build great relationships with customers. In a recent doorstep interview on BBC news, one Abel & Cole customer admitted that she couldn't ever cancel her order because her seven-year-old daughter was in love with their delivery driver.

Practical service

Abel & Cole's aim of leaving its customers feeling happy has been translated into a range of practical service provisions that aim to make it easy for people to get what they need from the business. Key provisions include:

- **Regular deliveries:** the flexible, regular delivery model developed for the company's core range of boxes has been extended to every item it sells. Each item in the catalogue can be ordered for any frequency, from once to every week, every month or even every twelve weeks. This means that customers can set up their shopping programme once and then do nothing. For example, an order could include butter every week, but rotate around four different types of bread, so that a different loaf arrived every week. Washing-up liquid might be set up to appear every four weeks and perhaps a case of beer or wine every couple of months. Customers tend to take a lot of time and care over the initial order set-up and then forget the precise details of what they've done. The result is that there is an element of pleasant surprise when the order arrives each week, especially when a new case of wine appears just as the last bottle runs out. As one customer put it:

'There's a certain childlike, Christmas-morning excitement about coming home, retrieving the box from the shed & rummaging through it to see what's inside.'(Abel & Cole email, 9 February 2006)

- **Easy web shopping:** customers can order anything from the range, with any frequency of delivery, on the website. They can also set their preferences for the exclusions service, alter delivery instructions, change payment details, review past orders, obtain a statement of account, etc. The website now accounts for 80% of sales by value.
- **Dedicated drivers:** unlike many home delivery companies, Abel & Cole does its own deliveries rather than using a courier. Furthermore, it allocates a dedicated driver to each customer, so each customer knows the name of their driver. Many have sufficient trust in their driver to give him/her keys to their house so that the delivery can be put in the kitchen. This also gives drivers the opportunity to get to know their customers, which adds to the customer relationship.
- **Deliveries while customers are out:** one of the major downsides of home shopping is the requirement to wait at home for the delivery. Typically, if one is not there to receive a courier delivery, it will be deposited in a central depot from where it must be recovered, an enterprise that can at times require a fair degree of determina-tion. Abel & Cole overcomes this issue by taking very detailed instructions from each customer about where the delivery should be left if they are not at home to receive it. With a bit of ingenuity, there is almost always somewhere it can go; if not a place in the garden, then a neighbour, a local shop or tucked away in the communal hallway. New customers are sometimes a little apprehensive about these arrangements, worrying about what will happen if the delivery is stolen from its hiding place. Abel & Cole reassures them with a delivery guarantee: if they call to say that the delivery was stolen, they will be refunded for the value of the delivery. No proof is required; the customer's word is enough. In practice, very few deliveries are stolen, and even fewer customers have abused the guarantee.
- **Automatic payment:** Abel & Cole takes payment from all its customers by credit card. Crucially, this avoids the need for them to pay the driver in person. This

protects drivers and means that there are very few credit control issues to deal with.

* **Weekly newsletter:** Abel & Cole includes a weekly newsletter with every delivery. One of its roles is to help the customer enjoy the produce in the selections: it includes preparation ideas, storage tips and recipes for any tricky fruit and vegetables that have been sent out. The website also offers recipes, organised by ingredient, which are very helpful for customers who may never have seen some of the items in their box before.

Most features of the service were developed in response to conversations with people who decided not to set up an order or who stopped ordering, for example, because they were at work all day and couldn't take a delivery, or because they weren't sure how to cook the food in their box, or because they were fed up with writing cheques to the driver every week. Each key aspect of the service was a direct response to a problem for a customer.

Sometimes, the customer with the problem was also an Abel & Cole staff member. The directors of the company have received a delivery every week without fail for several years, and their experiences have been a continual source of ideas for improvement. Often, it is the smaller details of the delivery that a customer might not mention which are picked up by staff and directors. For example, Abel & Cole staff used to joke about how difficult it was to find somewhere to keep their box for return to the driver; this led to the design of a fold-down box that fits easily under the sink or stairs.

With the exception of the website, all of the key aspects of Abel & Cole's service were in place by 1999, paving the way for the broad appeal and rapid growth of the business thereafter.

Sourcing

Abel & Cole has sourcing policies that extend well beyond organic production. The aim is to ensure that every aspect of procurement is aligned with the values of organic production. As a minimum, nothing should undermine or contradict the ethical and environmental benefits of the organic production. For example, the company would not buy from an organic producer who didn't respect workers' basic rights. Wherever possible working practices should actually further the benefits of organic production in a positive way. For example, the company buys fairly traded organic produce whenever possible. This approach to buying has been in place for many years and started informally, with Keith Abel doing the buying in a way that felt right to him. Over time, the right way to buy has been formalised into a set of buying policies and practices.

Abel & Cole's comprehensive approach to ethical procurement is highly valued by most of its customers. For many, it is the primary motivation in their decision to buy from Abel & Cole. Different aspects of Abel & Cole's buying policies appear to have particular relevance to different customers. The most important aspects are discussed below, followed by an explanation of how Abel & Cole has implemented them.

Production

Where an organic standard exists, Abel & Cole will buy only goods that carry that standard. Over 90% of Abel & Cole's range is organic, including all fruit, vegetables, meat, dairy and pantry goods. Abel & Cole does include in its range some goods that cannot be organically produced because they are not farmed or made from farmed ingredients – for example water, wild foods, and cleaning products. In these cases, Abel & Cole bases its purchasing decisions on its own research into the environmental issues relating to the products and the credentials of potential suppliers.

Location

Abel & Cole's policy is to buy produce as locally as possible. This policy provides the nation's farmers with support; it reduces food miles (and therefore pollution); it means the food is fresher and it keeps people in touch with the delights of each season. Anything that is available from British producers will be bought from them. This means that 100% of the company's meat, fish and dairy ranges are from British producers. Beyond this, the company actively seeks producers who are as close as possible to their customers. Thus, for example, three different organic bakeries, located in south London, north London and Long Crichel, Dorset, bake bread daily for delivery to the customers nearest them. This enables the company to ensure that every customer's bread is baked within 60 miles of where they live.

For fruit and vegetables, the ability to source locally is highly dependent on the seasons. During the peak of the season, 100% of Abel & Cole's produce is British. Taking the year as a whole, Abel & Cole's import level for fresh fruit and vegetables stands at around 30%, significantly lower than the major multiple retailers. Some imports are required to provide boxes throughout the year, particularly if those boxes are to offer some variety and nutritional balance. In particular, domestic fruit production has a very limited range and season.

Producer relationships

Abel & Cole's policy is to buy directly from the producer wherever possible. Removing the margin taken by agents and middle men is of financial benefit to both the supplier and Abel & Cole. In addition, working directly with the grower enables a good exchange of information, quick resolution of problems, and conversations that frequently lead to new, mutually beneficial ideas. Abel & Cole does not have a fixed approach to the way it works with growers. It aims to be flexible and pragmatic, developing collaboration and consultation at a speed dictated by the grower.

Despite the fact that working without a middle man is of mutual financial and practical benefit, Abel & Cole has had to work hard to persuade growers to deal directly. In the early days, when the company was very small, growers perceived it as high risk. They felt that Abel & Cole might go out of business, leaving their bills unpaid. Given the large number of start-ups that do go out of business, this was not an unreasonable fear. The small size of the company also translated into small purchasing volumes; growers were reluctant to invest time in selling directly to Abel & Cole when it would be able to take only a small amount from them, and when the same time investment with a different buyer could lead to the sale of their whole crop. As Abel & Cole has grown, these issues have receded, and buying directly has become easier.

Price

Abel & Cole offers to agree volumes and prices in advance with fresh produce growers before the seeds have even been purchased or planted. Once again, this approach has been taken because seems like a reasonable thing to do. If Keith Abel had come from a professional buying background, he might have acted differently. Most produce buyers in Britain work for the multiples or the packhouses who supply them. Demand and prices paid by the multiples changes from day to day according to what is selling well in the stores. Most growers therefore plant and cultivate their crops with no certainty about how much they will sell it for, and indeed whether they will sell it at all. There are no written contracts in the fresh produce business, so even once a grower has taken an order and harvested his crop to meet it, he or she cannot be completely certain that the agreement will be honoured.

Despite the high levels of risk and uncertainty present in the standard way of working, not every producer wishes to accept Abel & Cole's offer of fixed prices. Some of them prefer to take the chance that the market price at the time of harvest (the 'spot' price) will be higher than any price they can pre-agree with Abel & Cole. The prices that Abel & Cole pre-agrees tend to fall between the high and low price points in the spot market, however, so the gains a grower could make selling at the spot price one year can be offset by losses another year. Therefore, most growers find that working with a fixed price is more or less equivalent in the long run. They have an incentive to honour the agreement in years where the spot price is high because doing so reinforces the relationship with Abel & Cole, which safeguards the opportunity to sell at above the spot price in future years.

Transport

The policy of buying food as locally as possible means that 'food miles' are kept to a minimum, and with them the pollution generated by transportation. When produce is not available locally, Abel & Cole has a strict no air freight policy, which means goods must be available for land or sea transport if they are to be sold by the company. This policy is a response to the high level of pollution generated by air transport, something that the company feels undermines the value of organic production. The no air freight policy has a high cost in terms of lost sales and inter-rupted orders. For several years, the policy was quietly pursued despite its costs. When air freighting became a matter of concern for consumers in 2003, Abel & Cole was in a very strong position, and perhaps made some gains that compensated for the costs of the policy it had put in place five years before.

Policy development

The fundamental goal of Abel & Cole's buying policies – to work in a way that is consistent with the values of organic production – are firmly established. The poli-cies that are followed to meet this goal, on the other hand, are by no means fixed; indeed they are subject to frequent debate. Discussion covers everything from big political questions (should the company sell Israeli produce?) through to matters of fine detail (are fair trade stickers a waste of resources or a good way to promote the standard?). Questions are raised by the buying team, but also by customers, the marketing team and members of staff. Often such questions require careful research before discussion and decisions can happen. What looks more ethical sometimes isn't; for example, glass bottles may feel more 'green', but are only better than plastic in environmental terms if they are collected back, washed and reused several times.

The buying team

Abel & Cole has encountered great difficulty in finding the right people to undertake its buying. The character and motive of the buyer is vital, with an ability to listen to growers being essential. It was also felt that professional buying skills were needed to handle big negotiations and to identify technically competent growers and inspire them with the confidence to supply the business. The problem was that the great majority of people with these skills had learned them in the mainstream produce market, where they had also acquired an approach to growers (and in some cases to Abel & Cole itself), which was at odds with the values of the business. After some unsuccessful and expensive episodes with professional buyers, Abel & Cole decided to develop its own buyers instead. A buying team was created largely through internal promotion of staff with the personal qualities required: perseverance, integrity and an interest in farming. This approach quickly proved to be successful.

Ethics

The aim of upholding the ethical and environmental values of organic food production extends beyond Abel & Cole's procurement policies into every aspect of running the business. This approach started with the founders of the business. Over time, the company has attracted staff who share these values and further them in the decisions that they make. Similarly, the company has attracted customers who share these values and expect to see them upheld by the company in everything that it does.

Running an ethical business

Abel & Cole does not have a simple system or principle that it follows to ensure that it lives up to the ethical credentials of its products and to the expectations of its customers. Success in this requires people throughout the business to consider environmental and ethical factors in all of their decisions. The large number of decisions made in the business every day renders it impossible for one person to make them all, check them all or serve as the conscience of the business. Furthermore, the variety and complexity of decisions involved in the running of the business means that they cannot all be covered by a simple set of guidelines to which people can refer. Ultimately the business depends on its staff to make their own decisions ethically. The personal qualities that will motivate and enable a person to do this are a high priority in the selection of staff at Abel & Cole. The more senior the staff, the more influential the decisions they will be taking, and therefore the more closely these attributes are examined. Abel & Cole has developed its own recruitment methods to help bring out these qualities in its management candidates. It has also invested in building peer support groups so that people can help one another with difficult decisions.

Whilst managers may take the final decisions about big environmental steps forward, such as eliminating air-freighting or introducing biodiesel, the opportunity to conserve resources is present wherever resources are used, and most of them are used at the shop floor level. Abel & Cole is fortunate to attract many employees who share its environmental values already; many more are converted by their co-workers. The commitment of its staff is essential to the company's success in packaging reuse and waste reduction. Staff have also taken initiatives to procure everything from coffee to electricity ethically.

Abel & Cole has treated good management practice as a central ethical requirement. It has implemented HR best practices, and has developed management approaches to give staff security, support and involvement – they have been ranked as one of the UK's top 20 workplaces.[3] With these vital foundations in place, Abel & Cole has also given attention to local recruitment, and welfare projects such as cycle schemes and English lessons. The company is motivated to take care of its staff because doing so is part of living up to the values of organic production. It also produces many benefits for the business, including staff who are sufficiently happy and motivated to provide exceptional customer service and to look for their own ways to further the environmental and ethical aims of the business.

The benefit of working ethically

One motivation for Abel & Cole to ensure that ethics run right through the business is to ensure that customers are never disappointed by anything they discover about it. Failure on this front would of course be at odds with the goal of leaving the customer feeling happy. It could also lead to significant loss of business. As ethical considerations motivate many customers to buy from Abel & Cole, many people would be motivated to leave if they lost faith in the company's integrity. Living up to people's expectations is therefore essential for the health of the business. At times this can be difficult: expectations exist that are not created by the company and are not realistic, for example, the expectation among some customers that their fruit and vegetables will be 100% British all year round.

While many of Abel & Cole's decisions need to be ethical in order to meet customer expectations, other decisions are invisible to the customer. Abel & Cole nevertheless seeks to ensure that these too are made ethically. Examples include the decision not to use air freight, well before consumers were aware of the issue. Smaller examples could include helping an employee who is failing in their work to retrain, rather than dismissing them, or giving somebody who is struggling with their work–life balance some paid time off. Some of these 'hidden' ethical decisions may in any case eventually come to light. A failure to take the right decision on something like air freight before it became public could have been very damaging. The business cannot, however, control or predict which decisions will come to light. Ethical consumers are highly motivated to find out the truth and tend to have the resources to do so (education on the issues, web information, newspaper readership). Revelations of hypocrisy from people presenting themselves as highly ethical is also perfect material for the newspapers, so they too watch ethical businesses carefully. If a business that presents itself as ethical makes an unethical decision, it may well be discovered and that information circulated quickly, undermining faith in the business, with potentially disastrous effects.

Some decisions may never come to light, yet Abel & Cole still aims to do the right thing. It believes that unless those decisions are also made correctly, the company will lose the faith of the staff who witness them, and they may in turn lose their energy to engage well with the customers and work hard for the business. Abel & Cole therefore believes that to achieve lasting success in the ethical marketplace, a business must be genuinely motivated to make ethical decisions at all points, not just where the customer can see them.

Abel & Cole's success has been driven by the fact that an increasing number of consumers are following their principles in their shopping decisions, and that for many people those principles extend beyond buying an ethical product (in this case organic), to include a requirement that it is sold in a way that is consistent with that product. Consumers who follow their principles beyond the product and into the broader picture are sufficiently motivated to seek out and pay for a product with an ethical supply chain. They are also sufficiently motivated to watch that supply chain carefully, and to hold the retailer accountable if their expectations aren't met. So far, Abel & Cole's own interest in creating an ethical supply chain and an ethical business has satisfied – perhaps even delighted – many ethical food shoppers. Its future success depends on its ability to continue to meet their high standards.

References

1 Williamson, S. and Cleeton, J. (2005) *Soil Association Organic Market Report*. Soil Association, Bristol.
2 Mickey, A. (2006) *Abel & Cole Newsletter*, Week Commencing 23rd January. Abel & Cole, London.
3 The Great Place To Work® Institute (2006) *Abel & Cole Feedback Report*. The Great Place To Work® Institute, London.

Chapter 7

Case History: Clipper Teas

Lorraine Brehme
Director and Co-founder, Clipper Teas

> Whosever commands the trade of the world,
> Commands the riches of the world
> And hence the world itself
>
> Sir Walter Raleigh 1552–1618

Introduction

Clipper Teas was started by Michael and Lorraine Brehme (Figure 7.1) in 1984 with £50 and two chests of single estate tea. Mike had trained as a tea taster, buyer and blender and Lorraine's background was in the health food trade.

From the humble beginnings of Mike buying and packing all the tea and Lorraine doing the sales and marketing, Clipper now employs 80 people at their factory and offices in Dorset and is sold in more than 30 countries, including America, Australia, Russia and Japan. The current ranges now include over 100 different teas, drinking chocolates and coffee in both roast and ground and instant formats. Most products carry either the Fairtrade Mark or the Soil Association Organic Mark, and in many cases, Clipper products carry both.

Figure 7.1   Mike and Lorraine Brehme.

Challenging conventions

When tea was first brought to the West it was green and was originally sold in apothecaries as an aid to digestion and for various other stomach ailments. By the 1930s, most tea sold in the UK was a blend of black teas from Sri Lanka (Ceylon) and India. This gave a rich and refreshing cup of tea. Lack of investment, low tea prices and a food shortage in 1973, however, brought the Sri Lankan tea industry to crisis point. The plight of the tea estate workers and the squalid conditions in which they lived was shown to the British public by television programmes such as *World in Action*. In 1974, the Sri Lankan Government nationalised the bulk of tea estates as part of a policy of land reform, resulting in the departure of the majority of large tea companies, taking mass tea production to Africa and South America. This completely changed the taste of the great British cuppa: like wine, the taste of tea is dependent on where and how it is grown and on climate.

Speciality teas were also available in the UK but if you bought a Darjeeling or Ceylon, frequently there was small type which said 'blended with other fine teas'. We knew the trade was missing the point and believed we could offer something better. Clipper introduced single estate teas, literally unblended teas from individual tea gardens, which were sold into the health food or wholefood trade. Potential large competitors had not been successful at selling through wholefood stores at the time as both tea and coffee were perceived as 'bad for your health'. Our ethical company philosophy, the quality of the teas and the innovative packaging presentation appealed, however, and Clipper became one of the first tea brands to have a nationwide distribution network via this route (Figure 7.2).

Thinking outside the basket

Along with bread and milk, tea had been perceived in the UK as a low value 'shopping basket' item. It also has among the most brand-loyal customers. People drank the brand of tea they were brought up with. Twenty years ago customers in the aisles at the supermarket browsed the many different varieties and brands in the coffee section, taking time over their selection, but at the tea section they would go straight to their usual brand, ignoring all the others. Also, the tea section was possibly the least interesting aisle in the store, with the mainstream brands dominating the fixture, many facings wide and the speciality section or fixture was small in comparison with only single product facings. How could a small company like Clipper change these entrenched patterns?

Late in 1991, we saw a small reference to the then embryonic Fairtrade Foundation in a book about the vagaries of global business. To have an independent organisation recognise and monitor the way businesses conducted themselves when dealing with Third World producers looked like the way forward for Clipper.

Figure 7.2   Clipper's first tea leaflet c.1987.

It is important to point out here that tea estates are often whole communities. There can be several thousand people totally dependent on that one estate for their livelihood, housing, the education of their children and the health care of the whole family, similar to conditions on country estates in the UK 150 years ago. If you had a benign caring employer, life could be good, if not it could be desperate and short. If you were born on a tea estate, you would most probably live and die there.

Clipper contacted the Fairtrade Foundation, told them of our interest in what they were setting up and were invited to get involved. We met with the founders who explained that they had fair trade standards and inspection criteria for coffee but that was all. Coffee and tea are grown and processed very differently. With coffee it is the seed (bean) of the plant that is used, with the best coffee grown by independent smallholder farmers and their families, rather than large plantations. The Fairtrade Foundation only had inspection criteria and established standards for smallholder coffee farmers and had no idea how to set criteria for tea, where the best quality is grown on larger estates that can support thousands of people. As it is the tea leaf that is used, it is imperative that it is handled carefully from the start of the process to the end, otherwise the end result is a bitter, unpleasant brew. It is generally a more sophisticated process to produce a good cup of tea than it is coffee and requires a great deal more infrastructure and organisation.

In the autumn of 1992, we became official tea advisors to the Fairtrade Foundation, helping to set the criteria for large tea estates and setting the social premium payment. This was rather a complex task as the perceived value of tea in the UK was far lower than in the rest of Europe, with the exception of Holland, with coffee commanding a much higher price here and in Holland, but not in the rest of Europe. After about nine months we all felt we had a workable and fool-proof set of criteria and a fair social premium, which was around 10% of the cost of the tea per kilo. This social premium was to be paid separately into an account that was to be managed solely by an elected 'joint body' of tea workers. This 'joint body' was to be elected by the other tea workers and was to represent them fairly, taking into consideration the ratio of men to women living and working on the tea estate. This was a radical move, as women in most of the developing world have very little influence over their lives and are afraid to speak up in front of the men. It is important to note that this separate joint body account currently only applies to large privately run tea estates. The criteria also included banning the use of the top ten most noxious chemicals such as DDT and paraquat, which are still widely in use in the south of India.

Going organic – the total solution

For several years prior to working with the Fairtrade Foundation, we had been trying to find organic teas. This proved frustrating as there were so few companies growing organic teas at the time. Buying small amounts was expensive – at least three to four times the cost of conventional teas, which meant Clipper was only able to sell the teas on a very small scale. Most of the organic teas we found were from small-scale farmers and were organic by default, i.e. they could not afford the chemical fertilisers or they were from Darjeeling and at a cost far higher than UK consumers would pay. Persuading new producers to start cultivating tea organically was going to take some time and would probably take years. From using chemical sprays to obtaining organic status takes three years of cultivation and during that time, yields fall dramatically.

The chemical inputs in the form of fertilisers, herbicides and pesticides kill the bacteria and life in the soil that is necessary to break down natural fertilisers. The soil quite literally dies, there are no worms and very little, if any natural activity, and so the plants become completely dependent on chemicals. Take these away and the whole 'ecosystem' simply stops. Those few tea estates that have made a success of changing to organic cultivation have one thing in common – a Planter (manager) who is completely dedicated to the change and ready to rediscover the cultivation methods used 80 years earlier. Old books were pulled out of dusty archives and traditional ways were combined with the latest understanding of how tea grows with nature. Large worms were introduced from Africa to India using a new system discovered by the Germans called vermiculture (worm culture). The worms were let loose to aerate the soil and bring the natural system back to life.

Burnside, Stockholm and the first fair trade tea

In 1993, one of the major UK tea brands planned to launch the Fairtrade Mark on one of their packs. The tea trade as a whole was very nervous of the Fairtrade Foundation at the time, however, and perceived those involved as unsympathetic trouble makers who were looking to challenge the status quo. At the last minute, the major brand changed its mind and pulled out. A lot of work had been done by the Foundation and its partners researching and setting criteria to enable tea estates to qualify for fair trade status and it was a real setback – and so Clipper decided to do it alone. The search was on for teas suitable for the Clipper brand and for producers who would be willing to meet the challenge of working with the Fairtrade Foundation.

As many smaller companies as Clipper could find were contacted and asked for tea samples and information about pay and conditions on their estates. Finally, we heard from a company in South India who exceeded our expectations with a black large leaf tea that had a wonderful flowery aroma and taste. They had taken over the Burnside tea garden three years earlier and had improved welfare standards dramatically in that time for the tea workers, and along with responsible management of the local environment, this had led to an improved quality in the tea.

Clipper applied to become Fairtrade Foundation licensees and they organised an inspection in 1993 of the Burnside tea estate, located in the Nilgiri Hills in South India. Burnside passed the inspection and became the first tea estate on the fair trade register of approved tea producers: we had the first fair trade tea we could sell under the Clipper brand.

As a small company with no corporate backing we were, however, taking a huge risk. Would the tea be overpriced in the general marketplace? We were still selling through the health-food trade and sales were growing, but we needed to also sell to the supermarkets to create sufficient sales to make the business viable. If we managed to get into the supermarkets, how would this affect our sales through the health food trade? Were there enough consumers out there who cared about where their food came from and the conditions under which it was produced?

The health-food trade by their very nature embraced this new thinking, but sales of tea through this channel were tiny in comparison to the millions of kilos of tea sold through the mainstream supermarkets. No new tea company had managed to take any shelf space from the large tea brands that continually dominated the UK market, and consumer demand for tea was slowly falling.

Our largest customer at that time was Community Foods who agreed to support us in launching Fairtrade Marked teas. Clipper had first started trading with Community early in 1987 as they were the largest wholefood wholesalers in the UK. Soon afterwards they became our distributor for that trade, which substantially increased our wholesale and retail customer base.

Launching change

The Fairtrade Foundation was officially launched in the spring of 1994. At that same time Clipper visited Burnside (Figure 7.3) and the company who ran it in South India, as well as another tea producer managing the Stockholm estate in the Dimbula district in Sri Lanka who were also interested in applying for fair trade

Figure 7.3   The Burnside tea garden.

status. Although passed by the FT inspector for satisfying the strict criteria, we needed to see for ourselves what was happening on these two estates.

The tea trade is notorious for highly disadvantageous trading terms for producers in southern countries and India is no exception. Low prices for the crop and lack of funds and investment are some of the problems facing the tea industry, which in turn often leads to exploitation of the workforce. Since independence in 1947, legislation has contributed to improving conditions for tea workers in general, but legal requirements are often ignored: tea estates are generally in remote, sometimes mountainous areas and many tea workers live in appalling deprivation.

Burnside is situated 8000 feet in the Nilgiri Hills and is surrounded by virgin forest, which is also home to panthers, leopards, monkeys, bison, boar, king cobras, different species of deer and eagles, and most recently, a tiger.

One tea worker on Burnside, Shivagami, told us that she was taken out of school at the age of eight to look after her younger siblings. At twelve she joined her parents as a construction labourer on the roads. At nineteen she married and went to join her husband on this tea estate, two weeks later she started work as a tea picker. At that time the estate was owned by a British company, the work was dawn until dusk, there was no employment contract or maternity or child care and wages were very poor, with women's wages paid to their husbands. Housing and sanitation were minimal and women were encouraged to have children to keep up the workforce.

In 1991, the estate was bought by the present owners, a South Indian tea company who have been developing very high welfare standards on their Nilgiri tea estates since buying their first estate over 30 years ago. For Shivagami's community, life changed dramatically (Figure 7.4). All tea workers, both men and women, were given employment contracts and no-one under the age of sixteen is employed. There was a set working week, with bonus payment schemes, pay has been trebled and no women's wages are paid to their husbands. There was proper maternity and child care, education in family planning, health and hygiene, paid maternity leave and properly maintained crèches for pre-school children. Families were given more land for growing their own food and keeping cattle. Every worker from the field to the factory is educated about the whole tea process, so they feel more involved in the end product. Everyone living on the estate was educated about their local environment, the need to conserve the natural forest habitat around them and the animals living within it. No-one now has to leave the estate on retirement when there are no younger family members still working on it. All children were given a good, free education on the estate with real opportunities for further education not only at university, but also at the recently set up vocational college. Although legally education in India is up to age fourteen, children on all these estates are now given the opportunity to stay on until sixteen and then go on to university or the vocational college, which has been recently built with joint funds from the fair trade social premiums, which are then matched by the estate owners. Women also have equal opportunity for promotion within the tea estate.

Figure 7.4   Shivagami and some of her fellow co-workers.

For Shivagami and her co-workers, life has changed for the better, although it is still hard by our standards. Shivagami's day starts around 4.00AM with family chores, she works picking tea from around 8.00AM until 4.00PM, then goes home to collect firewood, water, cook the family meal, etc. The difference that matters most to Shivagami and the other parents on the Fairtrade Mark estates is that their children now have the opportunity to change their lives, to become doctors, engineers, teachers or any profession they choose.

These estates also protect the virgin forest that surrounds them, keeping away illegal loggers and poachers. The owners found that by looking after their workers and educating them about what they were doing and their environment, an unforeseen effect was that both the quality and the output of the tea increased.

In Sri Lanka it was a different story. Although conditions for the workers are better than in 1974, there has not been a dramatic improvement in the position of the tea pickers. Since 1992 the State has been selling management contracts to private companies in an attempt to remedy the lack of investment. It was one of these private companies, also Sri Lankan, which had taken over Stockholm (Figure 7.5). They also believed that the tea workers should get a fair deal for their labour and that their children should receive a proper education. At the time Stockholm had 816 employees; including their families meant that around 5000 people were living, working and totally dependent on the estate.

Figure 7.5   The Stockholm tea garden.

Clipper launched the first tea in the world to have a Fairtrade Mark in November 1994 in the Houses of Parliament in London (Figure 7.6). Organised by the Fairtrade Foundation, there were several eminent MPs present and several representatives from the national press. Over 100 MPs signed an Early Day Motion that Fairtrade Mark products should be available in Parliament and to this day Clipper tea is sold through the Parliamentary canteens.

The Clipper Fairtrade tea range was launched through larger Sainsbury's stores, delicatessens and the wholefood stores. As with our very first customers in 1984, once the tea buyers tried our tea they were amazed at how good it tasted, 'tea as it used to taste, light and refreshing' was the usual comment. It still took two years to get into all the main supermarkets, however, and even then it was mainly into 'selected' stores. Clipper has opened the door to mainstream sales for those new, smaller companies – the first new tea company in over 30 years to do so.

Into the mainstream

Also in 1994, we finally discovered a fantastic organic tea at a competitive price. It was a single garden, small leaf Assam from the oldest tea company there. Aware of a growing consumer interest in organic food and beverages and concerned about the

Figure 7.6   Clipper's first Fairtrade range of teas.

low cost of tea generally and the environmental impact from intensive monoculture, in 1990 they had decided that organic agriculture was the answer to both problems. Three years later they received organic status.

The impact on the Rembeng tea estate is that not only are the tea plants, wildlife and vegetation thriving and the tea workers healthier, the tea actually tastes better. The wonderful, rich, malty characteristics of Assam tea sought by connoisseurs, has returned. There are around 50 families, with between three and five family members living here. All workers have contracts and a set working week, leaving time for leisure in the recreation club provided where indoor games, reading materials and regular cultural programmes such as dancing, films and music take place. As Assam is one of the wettest places on earth, suitable wet weather clothing is provided to protect against the seemingly constant downpour and atmosphere of an overheated greenhouse. There are comprehensive medical facilities, schooling and maternity care is readily available and each family is given land to farm their own food. Decent wages are paid to both men and women and there is no child labour.

A unique feature of this company is that they set up the Rhino Foundation for Nature in 1994, established to protect the one-horned Indian rhino, of which there were then less than 1000. Their habitat is in neighbouring Kaziranga National Park. Kaziranga, an area of around 430 km², supports the largest number of rhinos on the subcontinent. Working in collaboration with the Forest Department and local non-governmental organisations (NGOs), the foundation has established a network among villagers living around wildlife sanctuaries, aimed at stopping poaching and reducing activities that harm the natural habitat. They also assist with the publication of conservation literature for distribution among the local population.

The first packs of Clipper Single Garden Organic Assam tea went onto the shelves in 1995 and won the Tea Category in the 1996 Organic Food Awards (Figure 7.7). It also won again in 2003. That tea estate went onto the fair trade register in 1997.

From the steady response to the fair trade range, located in the speciality fixture, it became obvious that a mainstream blended tea was needed to really optimise sales. In 1996 Clipper Gold was born, a blend of our Indian and Sri Lankan teas. Although in keeping with other Clipper packaging and sporting a large Fairtrade Mark symbol, we felt that it did not stand out sufficiently on-shelf to compete with the other mainstream brands, which could be ten facings wide to Clipper's one or

Figure 7.7   First Clipper Single Estate Organic Assam tea.

two facings. So, in 1997 we started a radical redesign programme and later that year introduced a range of organic 40 tea bags, a single garden Assam, Ceylon, Earl Grey and a blend that was launched through the health-food trade and several supermarket chains (Figure 7.8). In 1999, our organic 80 pack for the mainstream fixture was launched (Figure 7.9).

Because of the increasing sales of coffee and especially instant coffee in the UK, we felt the time was right for an organic version. In the spring of 1998 we launched Clipper Organic Instant Freeze Dried Coffee. As with all our products it had to taste as least as good as its non-organic opposition. We won Best New Product at the Organic Food Awards later that year and in 1999 we launched a Decaffeinated version and the first Organic Coffee Granules (Figure 7.10). The interest in organics was steadily growing and all of these coffees went into the supermarkets as well as the independent trade.

Clipper worked closely with coffee trading partners to get the right product and some of the smallholder coffee growers and their families were visited. These organic coffee 'gardens' are usually owned by one family with anything from 1.5 hectares to 20 hectares and set within their local environment, whether in the jungle in Costa Rica, the coastal mountain range of the Dominican Republic or along the border of the central highlands and the Amazon basin in Peru. For many of these families, the

Figure 7.8   First Clipper Organic Speciality tea range.

Figure 7.9   Clipper Organic 80 pack for mainstream markets.

Figure 7.10   World's first organic instant coffee.

method of organic farming is permaculture. This means that many different varieties of food that grow at different heights can be grown together as occurs naturally in the jungle, giving far higher yields. So a family can grow coffee and cacao as cash crops to pay for items such as clothing, medicines, schooling and sugar, along with all the vegetables and fruit while also keeping pigs, chickens and goats. They have plenty to eat and any surplus goes to market for extra cash. Ayuveda, the eastern herbal practice, has been introduced and is one of the main methods used for the organic control for plant disease.

In Peru, our coffee farmers earn more from growing organically than they were getting from the cocaine drug barons for growing coca, and they now live in peace rather than in fear of violence (Figure 7.11).

The Clipper brand emerges

As the organic and fair trade ranges grew with the addition of roast and ground coffee, mood teas in revolutionary packaging and herb teas, we realised we had been so busy increasing our supplier base, our product ranges and also making sure the packs were eye catching and different, that we had taken our eyes off the Clipper branding. This became evident one day walking down the tea aisle with the buyer of one of the top multiples. They stocked an enormous range of our products in the

Figure 7.11   Organic coffee farmers and crops in Peru.

mainstream, speciality and coffee sections and each individual range looked good. Except that you could not immediately identify them as a Clipper product, each range could have been a different brand as the branding had become small and the packs so busy. Talking to some of the customers it became evident that most didn't know we produced anything other than their favourite tipple. Our tea purchasers didn't realise we also had coffees and vice versa.

We needed to turn our attention to developing the Clipper brand image. We knew we had the right products at the right price but we still did not have the packaging to match. A brand-strengthening exercise was called for, so in 2000 we called in several design agencies to see what we could achieve. We chose London design agency Williams Murray Hamm to streamline the ranges and produce the simple but extremely effective packaging principles we still use today (Figures 7.12–7.15).

Figure 7.12   The award-winning Mood range.

Figure 7.13 New and current packaging style – Organic Herb tea range.

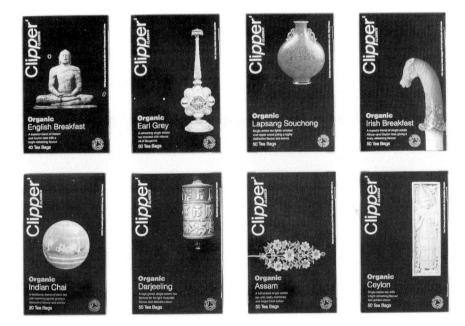

Figure 7.14 The Organic Speciality tea range.

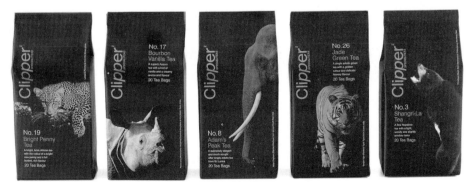

Figure 7.15 The multiple award-winning Specialiteas range.

Innovation

The publicity generated by the new designs saw sales soar and once again we wanted to bring more innovation to the beverages sector. Working with organic producers in remote regions of the world, soon shows that the West and its modern attitude to agriculture and medicine does not possess all the answers. Local indigenous plants were used time and again to combat pests and diseases in a variety of ways, especially as prevention. Ayurvedic practice could even help where modern chemicals had failed, as with banana blight in Peru.

We employed the services of a qualified Yogi who advised on the types of herbs for various types of people. We then experimented with different recipes to find teas that would be suitable for the British consumer.

The range of Clipper Ayurvedic teas was born (Figure 7.16) and according to many of our customers, the Detox is a wonderful hangover cure.

By the end of 2001 we had a range of single estate, unusual origin teas, each of which told a romantic travel story of that country, a range of Ayurvedic medicinal teas, a huge range of herb teas including nettle and dandelion, honeybush and redbush, a range of green teas, instant coffees and the most wonderful organic roast and ground coffee from Papua New Guinea. *The Grocer* magazine awarded this coffee a place in the top ten food products in 2000.

Clipper then went on to introduce the first organic instant cappuccino, a range of Cafebar type roast and ground coffees, a range of the first organic instant hot chocolates, the first children's tea for the UK and the first white tea range (Figure 7.17).

Figure 7.16   The Clipper Ayurvedic range.

Figure 7.17   Part of the extensive current Clipper range of teas, coffees and hot chocolates.

Proving the difference

To celebrate the tenth anniversary of the Fairtrade Mark in 2004, Clipper took a representative from one of the leading UK supermarket chains to Sri Lanka to see just what ten years of fair trade have achieved there.

We visited two very uniquely different and beautiful tea estates. First we went to the estate whose tea we had been selling under the Clipper brand with a Fairtrade Mark since 1994. It is a gruelling six hour car ride from Colombo to our destination in the Highlands. Often described as 'the world's biggest monoculture', the tea-covered hills with their elegant shade trees of silver oaks are one of the most romantically beautiful places on earth. In those ten years of fair trade, Clipper had been able to donate to the estate over £200 000 in premiums, on top of the price of the tea, which the estate manager proudly told us that that year he had beaten his own record for the sale price of tea at auction. As Clipper buy around 30% of the estates output, it is important to know they still get a good price elsewhere. As the majority of the tea we sell is in teabag format, we encourage the producers to pack and sell good large leaf tea to other markets to maximise their profits.

As we drove around the different gardens on the estate, it was fascinating to see the changes that had taken place in the last ten years due in large part to our premium payments. As well as the purchase of two new ambulances for emergencies (Figure 7.18), three new community halls had been built and furnished, holding around one

Figure 7.18   Stockholm committee members and the new ambulance.

hundred people in each. Chairs face a small stage where different ceremonies can take place and the intention is for the estate workers to hire the halls out to local families at a competitive price. All the profits earned go back into the fair trade premium account to be invested in future projects. At the back is a kitchen equipped with a water pump so there is no more need to collect water at the river with buckets.

Although hardly luxurious, the workers' cottages have been vastly improved and now have electricity for lighting, cooking and television. Around the cottages are vegetable plots so the workers can supplement their diets and also sell any surplus to market. They also keep cattle for milk, again any excess is sold on. The average wage for an estate worker is about £35 per year and although they receive many subsidies, the fair trade premiums represent a huge opportunity for substantial, life-changing improvements

We sat in on a Joint Committee meeting where the group of fifteen pickers, supervisors and managers decide how to invest the Fairtrade premiums. The estate owners have no access to this money, it has to go into the separate Joint Body account. The committee has 60% female members to represent the balance of females to males living on the estate. Ten years ago the women were afraid to speak out, now they argue their case in the meetings with confidence. As well as offering low interest loans to local families, repayable over two years, they want to build, equip and staff a computer training centre for children living on the estate as a supplement to their on-site education.

The role of estate manager is vital, as workers and managers are obliged to take equal responsibility for decisions relating to fair trade premiums. This revolution in communication between the two groups may be the most obvious benefit of fair trade on large privately owned estates. There are five trade unions represented here, yet the committee members readily admit they now think more about the wider benefits of proposed investment than what individual unions stand to gain.

The second estate we visited was organic but had also been put onto the fair trade register four years previously. In direct contrast to the long-established Stockholm nestled in the shadow of Adam's Peak, the Greenfield organic estate is anything but conventional. The Sri Lankan government had given the land to the company ten years earlier, thinking it worthless and impossible to cultivate as a successful tea garden. Huge boulders are strewn all over the slopes, which are alarmingly steep in places and command a stunning view from the 1542m elevation. Contrary to expectations, tea grows well around the boulders, their taproots probe very deeply and find moisture under the rocks so effectively they have even survived a three-month drought.

Those running this estate had never grown tea organically before they came here and had a very steep learning curve. Their realisation that they had to work with nature rather than focusing on higher yields above all else, came as a revelation to them and they are now committed converts to organic agriculture. The initial low yields and problems such as nitrogen fixation were a real headache at the start, but now the organic methods used are in line with the traditional way tea was grown

Due to the mountainous terrain, soil erosion is a real problem in Sri Lanka, therefore tea with its long taproots and commercial lifespan of over 70 years is the ideal crop. Like our smallholder organic coffee producers in South America, however, our organic tea producers have found that diversification is the key to success. The company here is a pioneer of what they call 'analogue forestry' (permaculture), which aims to replace the disappearing tracts of natural rainforest. By planting densely in small areas using trees (fruit, nut or otherwise) to create shade and vertical columns for plants to grow at different strata, many different crops can be produced in the same area – from peppercorns to coffee and papaya and from mangoes to lemons and pineapple.

Once again, it is a committed and enthusiastic estate manager who makes the difference. Here, homes for the tea pickers and their families are maintained to good standards and social welfare is paramount. During our visit we watched a patch of land being dug to build new homes, paid for directly by the company, not fair trade premiums. As Clipper premiums are now just starting to stream in, the Joint Body is considering how to best invest the funds, their proposals being very similar to those on Stockholm.

To prove the importance of a committed, caring manager to make things happen, at a neighbouring estate the investment decisions have not gone well. The committee had several million rupees available but union leaders objected to some of the proposed community projects and wanted cash instead. Part of the fair trade standards that producers must follow is that the premiums are invested to the benefit of all, including future generations. It is easy to see the appeal of money being given directly to workers, but the development agenda of fair trade does not allow that – benefits must be shared and long-term.

Our guest from London was impressed, not only with the changes on the Stockholm estate over the ten years and the incredible changes in attitude of those living and working there, but also with the dedication of those running both these estates to making a fabulous cup of tea at the same time as looking after their people and their environment. That supermarket chain soon went on to offer a comprehensive range of their own private label organic and fair trade products.

Marketing

Effective marketing is crucial if a new brand such as Clipper is to fight its way into an old established market and stay there. As a small company we had to find a way to communicate directly, not only with our current customers but also with those who we wanted to try Clipper products in the future. In the winter of 1996 we started *The Teapot Times* (Figure 7.19), a free quarterly magazine, published in-house and that was established to educate all our customers about the real meaning of fair trade and organic. Fair trade is not an easy concept to explain to people in general. It is difficult for consumers to understand the difference between charity (an amount of

Figure 7.19 One of the *Teapot Times* covers.

money per pack going back to the producer on a short-term basis) and trading fairly with independently audited producers in a sustained and long-term partnership.

Along with *The Teapot Times*, which also went to retail and wholesale buyers, good PR was important – from television food shows to magazine tea tastings, from Commonwealth Heads of Government meetings to trade and consumer shows. We imported 10 000 tea plants from India for a 'tea garden' at The Eden Project, working closely with them to make it successful, and our packaging is a good advertisement for the quality of our products.

Conclusion

Since 1984, Clipper has won 30 food awards, 12 design awards and 4 marketing/ business awards, testimony to the quality and presentation of our product range. While we continue to run Clipper to the highest standards of integrity, it is ultimately the high quality and taste of the Clipper products range that generate repeat sales from our loyal customer base, and encourage new customers to try our products for the first time.

Chapter 8

Case History: Duchy Originals

Petra Mihaljevich
PR and Communications Manager, Duchy Originals

Introduction

As one of the UK's leading organic food and drink companies, Duchy Originals was one of the pioneering brands in this country's organic food movement. From modest beginnings in the early 1990s, when the brand was established by HRH the Prince of Wales to demonstrate that value-added marketing could be achieved successfully in organic and sustainable food farming and production, the Duchy Originals range has grown to include over 200 products in 2006.

The blueprint for the business model was established in 1991 and this model is still true to the ideals of the brand and the business today. The first product, an Oaten biscuit made with wheat and oats from the prince's Home Farm at Highgrove, was launched in 1992. During the 1990s the brand established its position within the market and in 1999 Duchy Originals made its first profit. From a humble biscuit, Duchy Originals has now grown to be represented in over 20 food and drink categories.

The brand has become a successful example of how to make an organic enterprise commercially viable. Duchy Originals achieved a major milestone in 2004, generating its first annual £1m profit. A core part of the founding principles of the brand is that all profits are donated to the prince's charities and by the end of 2006–2007 nearly £7m will have been donated.

The Foundation seed

The initial seed for the development of Duchy Originals was planted in 1986 when HRH the Prince of Wales took a decision to adopt organic farming principles on his Highgrove estate, near Tetbury in Gloucestershire. He had long believed that organic farming was one of the most sustainable methods of husbandry, ensuring the protection of the environment, countryside and wildlife for future generations. The Highgrove estate, under the management of David Wilson, underwent a programme of organic conversion and was certified as fully organic by the Soil Association in 1992. The prince, in his role as heir to the throne, became one of the most high-profile advocates of the organic movement, at the time under much scrutiny from a sceptical public and media.

In parallel to this he addressed a farming conference in south-west England in 1987. Sympathising with the plight of fellow farmers, he sought to highlight that value added marketing of farm produce was the way to improve profit margins. The emphasis in agriculture had to change, he said, to improving quality rather than just producing vast yields. 'The packaging and presentation of farm produce and the importance of adding value, through processing, preparation, grading and also the introduction of new products in general should be foremost in our minds. Failure to add value to a product in some way implies that you forgo to add value, to improve your profit margin, and thereby hand the chance to someone else.'

It was based on this concept that the Prince sought to lead by example. An advisory group of leading media, advertising, market research, product development and business consultants were drawn together in early 1990 to explore ideas for a value added marketing initiative based on raw materials from the Duchy Home Farm at Highgrove and other organic farms.

The creation of a brand

The first product to come out of this work was a loaf of bread made using wheat and oats from the Highgrove estate that went on sale in a limited number of Tesco stores in 1990. The product itself wasn't quite right, although it offered a starting point for considering how products could be created to add value to farm produce.

From this experience it was established that creating a 'brand' was the best means for achieving the prince's objectives. The brand needed to represent all the prince's aspirations and would encapsulate what the Prince of Wales calls the 'virtuous circle'. First, it would support sustainable agriculture and ethical production to benefit the environment and soil health through a more sustainable approach to farming. Second, it would demonstrate the advantages of 'value added' marketing through the production of a premium range of products using the highest quality ingredients and expert production methods. Third, once the brand was established, it would help to generate profits for the prince's charities.

The umbrella brand would provide a measure of quality for any products produced under its name. It would help to encourage UK producers and manufacturers to aspire to higher standards and quality, particularly in agriculture. In turn, it would aim to reconnect people with the food on their plates and encourage them to consider where their food was coming from and how it was made. All products would be British made and have a distinctly British feel.

One of the most important characteristics of any product would be that it must be something the prince would himself eat or use, enjoy and recommend.

The circle of integrity – virtuous circle

To launch the project, a biscuit was proposed as the best pilot product to test the brand concept, to be made using the wheat and oats from the Home Farm at Highgrove. If the pilot was a success, the brand would then start on a small scale to establish the provenance of the project and build a profile. In time, as the brand established itself, it was anticipated that a wide range of product areas could be explored. As long as the product upheld the integrity of the brand and the philosophies of the Prince of Wales, the brand could be stretched across food, drink and even to those non-food products where sustainability could be demonstrated. A 'circle of integrity' (see Figure 8.1), based on the virtuous circle, formed the basic template from which products could be measured to ensure they upheld the highest standards.

Code name 'Duchy Originals'

The development, code named 'Duchy Originals' was born. 'Duchy' in reference to the provenance of the brand, affirming the link to HRH the Prince of Wales as the Duke of Cornwall and the ambition to source raw materials from the Duchy Home Farm and Duchy of Cornwall estates. 'Original' because it was such a unique brand and each product produced would aspire to be completely original. The visual manifestation of the brand took some time to perfect. However, it was ultimately decided that the crest of the Duchy of Cornwall, slightly modified, would be the brand mark.

Manufacturer model

It was anticipated that the best approach for Duchy Originals products was to use existing high-quality UK manufacturers or producers to make the products. In order to generate a profit for the Prince's charities, a royalty from sales would be agreed

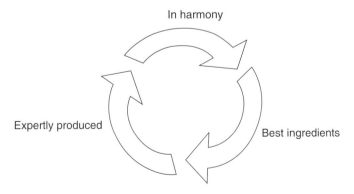

Figure 8.1   Circle of Integrity – Duchy Originals food production model.

with each manufacturing partner. The products could only be manufactured from the finest ingredients to a high specification set by the brand. The methods of production were regarded to be as important as the brand itself and quality was to be measured throughout the whole process and not just in the finished article.

In search of manufacturers, Chris Nadin, the first Managing Director of Duchy Originals, said they were seeking a company that was sympathetic to the philosophy of the brand and believed in what the prince was trying to achieve. Manufacturers also had to be the absolute best in their field with strong and experienced staff and a similar ethos to the Duchy Originals brand. Most importantly, they needed to be willing to produce organically. They did not need to be an exclusively organic producer, but they would need to be happy to conform to organic production standards and work closely with the Soil Association to ensure organic certification.

Renowned biscuit makers Walkers Shortbread in Aberlour-on-Spey in Scotland were recommended as the right company to work with. An independent firm with a huge reputation and a family business style, Walkers fitted the proposed 'Duchy Originals' ethos perfectly. James Walker, current Managing Director of Walkers Shortbread recalls the initial approach from Duchy Originals in 1991. He believes Walkers was selected to bake the biscuit, 'because they were big enough to cope and small enough to care.'

Baking the Oaten biscuit

Walkers perhaps did not anticipate the challenge they were set when they agreed to be involved. Chris Nadin recalls presenting 'a list of requirements as long as your arm'. 'The Duchy Originals biscuit needed to have "weight". To somehow convey in taste, texture and appearance that it was a serious biscuit – it had integrity, meaning, it stood for something. People had to take it seriously.'

One way of visually representing the brand's integrity was to mark each biscuit with an imprint of the Duchy brand mark. This was seen as integral, but presented its own challenges in the size, texture, shape and baking process. Each new batch would not only be taste tested but also tested with the mould to ensure it made the cut.

The biscuit needed to be made from natural and clean ingredients in accordance with its organic principles. The final recipe included only a handful of ingredients, yet creating the perfect balance ended up taking over 100 attempts to perfect. In terms of flavour, it needed to be hugely individual. The aim was to create a differentiated 'super digestive'; a biscuit that was suited to cheese and also sweet spreads or just on its own.

One part of the specification that was difficult to implement was that no hydrogenated fat could be used. At the time, there was little consumer awareness of hydrogenated fat and its health effects but removing it was a matter of principle.

On a practical level, removing the preservative from the biscuit severely reduced the shelf life of the product. The early product had only a three-month shelf-life, which caused complications and was a challenging issue to overcome.

The whole supply chain was instrumental in creating a point of difference. The organic oats from Home Farm had a unique flavour that set them apart from their conventional equivalents. The wheat called Maris Widgeon was a heritage variety well suited to organic production. The oats were stoneground at Shipton Mill near Tetbury to give them a rustic, coarse texture. The biscuits would be produced in small batches to preserve their integrity and then slow baked, similar to shortbread.

At each step in the process, the Prince of Wales was also involved in tasting the product to ensure it was right. Once the final recipe was agreed, the prince was also in attendance to commemorate the first official production run, visiting the Walkers bakery in Aberlour on 13 October 1992 (see Figures 8.2 and 8.3). The finished product was named the 'oaten' biscuit to give it the unique identity it deserved from the very beginning.

Figure 8.2   HRH the Prince of Wales inspects the first official production run of Duchy Originals Oaten biscuits at Walkers Shortbread factory in Aberlour on 13 October 1992. Photograph by Tony Spring. Reproduced with kind permission from the Elgin Museum.

Figure 8.3   HRH the Prince of Wales inspects the packaging of the oaten biscuit, Duchy Originals' first product at Walkers Shortbread factory in Aberlour 13 October 1992. Photograph by Tony Spring. Reproduced with kind permission from the Elgin Museum.

The visual brand

The look and feel of the packaging would be the physical manifestation of the brand. The packaging needed to have absolute integrity and convey the quality of the product within as well as the philosophies of the Prince of Wales. From an early stage it was agreed that there would be no commercial advertising or marketing of the products in the infancy of the brand. The Duchy Originals brand would rely on public relations coverage and word of mouth to build profile and sell the products. This meant the packaging needed to work overtime.

The packaging needed to be natural, smart and sophisticated. It had to be very stylish and classic so the design wouldn't date, understated but with a serious shelf presence. The whole design was crafted with the greatest attention to detail. Even the subtle shade of cream used was specifically chosen to give the packet a natural feel as opposed to a stark white. The front of the packaging was illustrated with an image of the oaten, simply stating 'Stoneground texture. Oven baked taste'. The side of the box showed very simple images of the biscuit positioned with its key ingredients, wheat and oats, on the one hand and the various serving suggestions, cheese and grapes, on the other. Everything was extremely simple, natural and clean.

The simplicity of the visual elements was complemented by the most important part of the packaging: telling the Duchy Originals story. Communicating the philosophy of the brand, the care and attention that had been paid in the production process and the provenance of ingredients was essential. Every element of the story endorsed the integrity and individuality of the Oaten biscuit and telling the story helped to reconnect the consumer with the origin of their food. The packaging described the establishment of the brand by HRH the Prince of Wales and the sourcing of oats and wheat harvested from the Home Farm, stating 'Duchy Originals is a coming together of expertise and craftsmanship'.

Distribution

The initial approach was to start the first phase of the brand's development on a small scale to establish the project. The scale was also limited by the supply of the raw ingredients from the summer harvest at the Duchy Home Farm, which meant the quantity of biscuits that could be produced was circumscribed.

One of the key objectives in launching Duchy Originals was to demonstrate the theory of 'value-added' marketing. The high quality of the organic ingredients and the expert production methods positioned them as the best in category and this commanded a price premium. Therefore the marketing of the biscuits was very much towards a niche premium market.

In line with this, the early distribution of the biscuit targeted small independent retailers with a focus on quality. Placement of the product was fundamental to getting the brand positioning right and was taken very seriously. The first 250 independent outlets in which the biscuit was sold were all vetted to ensure that they were right for the brand. If the products were a success and economies of scale could be established, creating a wider distribution network would be the next logical step.

The Oaten launch

The first official press announcement was issued on Wednesday, 4 November 1992. The tone of the release was fairly discreet, simply announcing, 'The first product of the agricultural marketing initiative set up under the umbrella of The Duchy of Cornwall was introduced today. Oaten biscuits, to be marketed under the new "Duchy Originals" brand, will be available from tomorrow from specialist grocery and delicatessens outlets throughout the United Kingdom.'

This still was, in essence, an 'experiment' and the success of the biscuit was yet to be measured by the critics. At the time of launch, organic was still seen as a niche concept. The prince was making a bold stand in his public support. More so, it was a unique move for a member of the royal family to be establishing a marketing brand, even if it was one with a very noble proposition. Inevitably there was negative press coverage.

It was the uniqueness and integrity of the product, however, that shone through. According to James Walker, 'There was nothing quite like it on the market at the time. It was a creative product that was genuinely different and that's why it worked'. The press coverage that was received following the launch recognised this.

Drew Smith from *Taste Magazine* said in the Christmas 1992 issue, 'These biscuits are so clean they are almost holy. They also taste pretty good… in the quest for integrity they have hit upon something that has character and stature.'[1] Joanna Blythman commented in *The Independent* in May 1993 that the Prince of Wales 'had taken the British biscuit and turned it upside down and inside out'. Blythman went on to say, 'one bite of a Duchy Original will convince even the most demanding consumer that the prince's biscuit amounts to more than royal hype.'[2]

1993–1999: brand development

The Oaten biscuit was followed by the launch of a gingered biscuit a year later, which received more excellent reviews. In the *Evening Standard*, November 1993, Henrietta Green commented on the success of the Duchy Originals biscuits saying, 'It is a shining example of what farmers and producers and growers can achieve.' It was the success of the biscuits that led the brand to explore further new product development – still operating on a relatively small scale.

The years between 1995 and 1999 could best be described as experimental for Duchy Originals. The brand was establishing its foothold in the market, building customer loyalty as well as exploring where the brand expansion could and could not go. It was a process of trial and error.

For instance, in March 1995, the brand launched its second wave of products: a range of fruit and herb adult soft drinks. The concept was unique and the drinks were positioned as a sophisticated alternative to alcohol. They were not organic but they contained no additives and preservatives. The raspberries were sourced from the royal estates in Sandringham and the elderflowers from the Duchy Home Farm. The decision was taken at the time to work with Callitheke, the owners of Aqua Libra and a subsidiary of Grandmet in conjunction with Coca-Cola Schweppes to produce the drinks. The drinks didn't follow in the footsteps of their successful predecessors. Quality was the imperative and the drinks just did not meet up to the high standards of the Duchy brand. It was clear that the Duchy Originals brand didn't work with big producers – the 'big enough to cope, small enough to care' model needed to be followed.

Further product expansion continued with the launch of a selection of specialty cheese, preserves, chocolates, tea and organic bread. Unfortunately, not all of the products launched during this period were organic, partially due to issues around raw material sourcing and a greater focus on the quality of production methods.

The biscuits, which were organic, remained the strong core of the brand. When the second edition of the *Organic Food Processing and Production Handbook* (2000)[3] was published, Duchy Originals gained a very brief mention for its biscuits but wasn't regarded as a major player in the organic industry at this stage.

1999–today: organic success

Finding the right balance of quality standards and organic principles was finally achieved under the chairmanship of Guy McCracken (ex-Marks & Spencer), who held the post from 1999 to 2004 and the current managing director, Belinda Gooding. The team of employees at Duchy Originals has grown from three in 1999 to over 20 today based in the company's head office in Richmond, Surrey.

Focus, quality, taste and integrity, based on the brand's founding principles, have been the key drivers in the company's success today. The Duchy Originals range now includes over 200 food, drink and non-food products. This growth has embraced the initial blueprint of the business in tandem with a strong product development, distribution and promotional strategy, which has led to the business establishing a strong foothold in the organic market in the UK. It has also achieved one of its ultimate objectives – to generate a profit for the prince's charities. Since 1999, the brand has gone from strength to strength, which is evident in current growth rate of around 24% year-on-year. In 2006–2007, the retail sales value of the brand will be around £56m with a profit of £1.2m (see Figure 8.4). By the end of the financial year 2006–2007, nearly £7m will have been donated to the prince's charities.

The brand has also been instrumental in helping to increase the profile of the organic movement in the UK. It has helped to support growth by creating a greater demand for organic raw materials and encouraging organic production methods to be adopted by UK manufacturers.

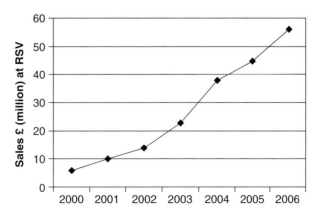

Figure 8.4   Duchy Originals growth in retail sales value for years ending 31 March 1999–2006.

The Prince of Wales has confounded those who didn't believe in his organic vision 15 years ago. His ambition to lead by example and create a successful example of 'value-added' marketing of sustainable produce has worked. The factors that have contributed to that success and whether they can be followed by other organic marketers is the focus of the second part of this chapter.

Environmental factors

The prince was ahead of his time in forecasting the rise in consumer awareness of food quality and desire for goods and services to be more environmentally sound. These issues didn't gain real momentum in the UK until almost ten years later. When the brand was established in 1990 it was seen as a fairly left-field concept. Yet by the turn of the millennium, consumers were searching for a brand just like Duchy Originals. These external factors have played a critical part in the brand's success.

The organic movement was still very niche at the time of the Duchy Originals launch. It was mainly spearheaded by the Soil Association along with pioneering brands such as Green & Black's and Baby Organix with a passionate focus on environmental sustainability and a drive for safer food. By the end of the 1990s there had been a significant organic boom in the UK, compounded by food scares such as bovine spongiform encephalitis (BSE), which generated consumer concern over conventional food production. Over this time, sensing the consumer mood, the supermarkets also increased their commitment to organic food. This critical mass helped to make organic raw materials more available, as well as making organic production a more attractive option to consider for producers and manufacturers to meet growing demand.

At the same time, running in parallel to this there was a growing movement from the mid 1980s to embrace high-quality food production and support British regional food culture. George Streatfield, Managing Director of Denhay Farms, Duchy Originals bacon producer, was involved in the development of Taste of the West in 1991. He recalls it being an 'uphill struggle' to promote high-quality regional and speciality British food at the time. These movements have also gained critical momentum from the rising consumer awareness and appreciation of food provenance and quality. The popularity of organisations such as Taste of the West and Slow Food today illustrate the resurgence of a strong British food culture with a focus on local, regional, quality production.

These two movements have come together in many respects and joined forces in a common drive for better food quality that continues to build strength with the consumer. Duchy Originals is a brand that sits at the crossroads, promoting the best of British in sourcing of organic and sustainable ingredients and 'adding value' through expert UK production.

Supporting market growth

The organic food boom in the UK has helped the brand to grow quite significantly since 1999. In the early days, raw-material sourcing was quite complicated. James Walker recalls sourcing organic ginger for the Duchy Originals gingered biscuits when they were launched in 1993. Organic ginger was originally sourced from China. It then needed to be processed within 48 hours of harvesting to preserve the flavour but the only organically certified processors were based in Italy. From China to Italy, the ginger then needed to be delivered to Walkers all within a relatively short space of time. According to James Walker 'it was very expensive and involved a huge amount of work.' Following the growth of the Duchy Originals brand, Walkers have been able to encourage their own ginger supplier in Australia to produce and process organic crops for the Duchy Originals biscuits.

The Oaten biscuit started out using ingredients from the Home Farm summer harvest. Today, the wheat and oats for the twelve varieties of biscuits in the Duchy Originals range still come from Home Farm and are supplemented by additional supplies from a much wider network of UK organic farms to meet demand. Organic butter for the biscuits was originally sourced from Austria when supply of organic dairy products was limited. Today it is mainly sourced from the UK.

Duchy Originals has also contributed to the increase in growth in the UK organic manufacturing sector in both expertise and employment. Like many of the Duchy Originals producers, Walkers Shortbread weren't producing organically when they were approached by the brand in 1991. In order to produce the biscuits, after achieving organic certification from the Soil Association, they had to establish networks with organic suppliers to overcome the complexities of organic sourcing and comply with organic production methods. Walkers now produce their own organic product lines, both within the Walkers Shortbread range and for own label clients. The Duchy Originals biscuits are by far the largest organic product range that Walkers produce, with up to 100 employees working to produce the Duchy Originals biscuits.

Crabtree & Evelyn, who produce the Duchy Originals preserves by hand, have been operating in Highbridge in Somerset for over 20 years. Since the launch of the Duchy Originals preserve range in 1998, the Duchy organic business now accounts for a majority share of the manufacturing that takes place in the factory. As with Walkers, Crabtree & Evelyn also now produce a range of organic preserves under their own label.

Unique selling points

The link to HRH the Prince of Wales and the brand's donation of profits to the prince's charities make Duchy Originals a unique business. The prince has been instrumental in raising awareness of the brand and making it a success. His beliefs and foresight have created a solid brand footprint. The prince adds integrity to the brand. This is from his desire to support sustainable farming and to inject a focus on quality not quantity back into British food production. His high standards in taste are also paramount as it has always been a prerequisite that any Duchy Originals product should be something the prince himself would eat and recommend. He takes a very active interest in the business and new product development work, approving all new products.

His interest is evident in the extension of Duchy Originals into other products in food and non-food areas. The range of garden tools launched in 2006, draw on his support of sustainable woodland management and UK craftsmanship. The tools are made using straight-grained ash from the Duchy of Cornwall woodland in Herefordshire, which is certified to Forestry Stewardship Council (FSC) standards for sustainable forestry management. The Duchy Garden tools are made by one of the UK's only remaining garden tool manufacturers, Caldwell's Manufacturing.

The brand has also been the vehicle to support the prince's ideas and philosophies on other pertinent issues. This is seen with the move into products using sustainably sourced Alaskan salmon and mackerel from Marine Stewardship Council certified fisheries (MSC) for a range of smoked salmon and fish pate. The MSC promotes environmental standards for sustainable and well-managed fishing practices.

The profit generated by Duchy Originals has been used to support a wide variety of organisations via the prince's charities. This support covers a broad range of areas including the build and natural environment, health, arts and education. Over £400 000 was donated to charities supporting farmers during the foot and mouth crisis. Some examples of the organisations that have been supported since the inception of Duchy Originals include Garden Organic (formerly HDRA), Elm Farm Research Centre, the Soil Association, Business in the Community's Rural Housing Project and the English Farming and Food Partnership, as well as international Red Cross appeals and a variety of other charities and hospices.

Product development

In order to make the brand a success and to generate a profit for the prince's charities, one arm of the Duchy Originals business strategy over the past six years has been focused on new product development to extend the range and stimulate growth.

In line with this strategy, the Duchy Originals range now covers a number of key food and drink categories, including:

- Grocery foods: biscuits, preserves, chocolates and condiments
- Speciality bread
- Meat: bacon, sausages and ham
- Ready meals and soups
- Refresher drinks and cordials
- Traditionally brewed ale

- Milk and dairy
- Seasonal poultry and ham
- Seasonal accompaniments

New product development is carefully considered and targets areas where the brand believes it can 'add value' to the category. The Duchy Originals ethos can expand into many different product areas as long as products adhere to the 'circle of integrity' model. It is important to identify areas where there is a genuine gap in the market where Duchy Originals products will have a real impact. Anticipating these opportunities has led to Duchy Originals 'raising the bar' in certain product areas to establish a demand for higher quality standards.

Biscuits and bakery products, including bread, contribute to a majority percentage of Duchy Originals sales (see Figure 8.5). Bread is a very strong performer, current production levels are at around 8000 loaves per day with an increase of around 30% on sales figures in 2005–2006 compared to the previous financial year. The meat categories are the next major product categories for Duchy, including bacon, sausages and ham. The Duchy Originals seasonal Christmas range now includes over 45 products and these additions to the range make the festive period the busiest time of year for the company by sales. This is contributed in large part by the suitability of the range to gifting and appeal to customers to trade up to more premium products over the Christmas period.

In addition to the strong core range, new additions such as fresh organic soup have been very well received by customers and the trade (see Figure 8.6). Chilled soup buyer for Waitrose Yseult Carioff-Richeux commented in *The Grocer* magazine in December 2005, 'Waitrose has a real following for organic soup. The arrival of Duchy Originals has driven strong growth with its four lines taking a big share of sales in Waitrose.'[4] In their first year of production, October 2004–2005, the volume of sales of the soup range increased three fold. The range, which retails at £2.89

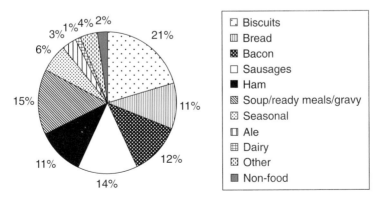

Figure 8.5 Percentage of Duchy Originals and Duchy Selections sales by product category.

a

Figure 8.6   [in parts a-d]: Pictures of the product range today. Duchy Originals Oaten Biscuits; Bacon; Beetroot Soup with Dill; Bread.

b

c

d

for 600 grams, has doubled to include eight varieties of fresh soup. The increased volume of sales has enabled distribution in supermarkets as well as independents such as Fresh & Wild. The move into the organic convenience sector has hit the spot with customers who are looking for easy eating options to suit busy lifestyles but with the added integrity of being organic.'

Duchy Selections – free range

Since 1999, the company has also produced a number of free range products, including bacon, ham and sausages. The pigs are reared to a very high Duchy Originals animal welfare standards on a par with their organic counterparts; however, they are not fed on organic feed or reared on certified organic land. Their feed is GM free to a high specification set by the company.

A number of other Duchy products cannot be certified organic. This includes the Duchy Selections Water, which is sourced from Pannanich Wells in Aberdeenshire near the Balmoral Estate. It also includes the range of sustainable fish products produced in conjunction with the Marine Stewardship Council (MSC). The MSC fish is sourced from the wild and therefore cannot be certified organic.

To avoid any confusion from customers and the trade about the non-organic items in the Duchy range a new brand name for these products was introduced in 2003. The Duchy Selections label is used for products to make a clear distinction to customers that they are not organic. Primarily Duchy Originals is still recognised as an organic brand and less than a fifth of Duchy products overall fall under the Duchy Selections label. The company continues to communicate the distinction between Duchy Selections and Duchy Originals, as well as the motivations for moving into these areas, using the packaging, their website and public relations.

Non-food development

The launch of a range of hair and body care products under the Duchy Collection label took place in 2004. The range is made using organic essential oils and natural plant extracts and includes a pure organic hard soap, certified by the Soil Association. The brand is currently working on developing production methods to make the range fully organic.

Duchy Home Farm and raw material sourcing

From initially sourcing the wheat and oats from the annual harvest, raw materials from Duchy Home Farm now contribute to around one-third of the organic product range of Duchy Originals. This includes oats and wheat for the biscuits, mustard seed for wholegrain mustard, parsnips, carrots and beetroot for vegetable crisps, milk from the herd of 180 Ayrshire cows and barley for the ale. Pigs for Duchy

Originals organic ham, bacon and sausages are reared on a nearby estate by farmer Richard Hazell.

The brand has always aspired, where possible, to source ingredients from the UK. All the meat and poultry used in Duchy Originals products is British. Admittedly it is not always possible to source all raw ingredients from Britain. This has become easier, however, with the growth of the organic market and Duchy Originals has been able to help support organic producers through generating demand for organic raw materials.

For instance, Duchy Originals was able to support the organic conversion of Worcester farmer Leslie James's orchard of damson plums to meet demand for ingredients for its damson preserve. The hand-picked Shropshire Prune variety of damsons are delivered to Crabtree & Evelyn each August. The preserve was recently awarded a bronze award in the best organic category in *The Daily Telegraph* Taste of Britain Awards 2005.

The prince is very keen to conserve the gene pool and preserve unique varieties of crops. This has helped to establish a point of difference for Duchy Originals products. For instance, the Duchy Originals ale is produced by Wychwood Brewery in Oxfordshire, using Plumage Archer barley from the Duchy Home Farm. The barley is an old variety that was originally used for commercial malting over 100 years ago but has since fallen out of favour over higher yielding crops with commercial breweries. The Maris Widgeon wheat used in the Duchy Originals biscuits is also a heritage variety, used in organic farming because it grows tall to help out space weeds.

Quality of manufacturing partners

Today, Duchy Originals works with a strong base of 30 manufacturing partners, including small to medium-sized businesses and small farmers and producers. All of the Duchy partners are located in the UK, a key requirement of the brand.

Producers range from the Amiss family in Devon who seasonally rear the plump Duchy Originals organic geese, to the Tracklements Company in Wiltshire who produce the handmade condiments, to larger scale producers such as Cranswick Country Foods in North Yorkshire who produce the Duchy Originals sausages. The network also extends to other organic producers in the supply chain, such as the Authentic Bread Company in Gloucestershire, who produce the Duchy Originals organic mincemeat for Christmas mince pies.

The company's brand managers work very closely with their manufacturing partners to set specifications and recipes for each product to ensure they have complete control of the brand. It is very important to recognise and credit these partnerships on the Duchy Originals packaging and each product has its manufacturer listed on the pack.

Ensuring absolute quality

Each product aims to be 'the absolute best' in its category, not just the best organic product in the category. This has become increasingly important as supermarkets have moved away from segregating organic products on the shelf – an organic product must be able to compete with its conventional counterparts. Ensuring the product is made with the best ingredients and the highest attention to detail in processing and packaging will help it to succeed commercially.

Adding value through the process

As with the Oaten biscuit, the processing of a product should always aspire to use methods that enhance its uniqueness, flavour and integrity. Adding value through the process of production is an important feature of all Duchy Originals products. This ensures never compromising on quality to cut corners or reduce costs.

The Duchy Originals organic Kelly's bronze turkey is a good example. The company approached Paul Kelly, based in Chelmsford, Essex, in 2000 to develop organic turkeys for the Duchy Originals seasonal range. Paul had a strong reputation for the quality of both his breeding and traditional processing techniques. At the time he was producing his own free range birds, he agreed to produce organic turkeys exclusively for Duchy Originals to Soil Association standards. The Kelly's bronze birds are a slow growing breed. Under organic standards, the turkeys are reared to strict animal welfare standards leading a free-range lifestyle on the organic land of a local farmer where they forage for worms and grubs supplemented with organic feed.

As well as having the best quality organically reared birds, the processing of the birds was integral to the finished product. The birds are traditionally dry plucked and then hung for 10–14 days before they are gutted. This adds a unique, almost gamey flavour and tender texture to the succulent meat. This is an expensive and labour intensive process and it takes some 50 extra staff each season to manage the Duchy birds alone. It greatly enhances the flavour of the bird, however, producing the very best organic Christmas turkey. Throughout the supply chain, selecting the right breed, rearing the birds to the highest standards and paying attention to the process adds considerable value to the end product.

Bacon is another example of how Duchy Originals has helped to 'raise the bar' in the organic meat category. In 1999 the brand began working with Denhay Farms in Dorset, who had a well-established reputation producing some of the best bacon in the industry and were renowned for their curing techniques. At the time, the conventional bacon market had been hugely devalued, compromising greatly on quality to cut costs and increase yields.

Denhay developed a special Duchy Originals dry cure involving sea salt, nitrite and sugar being rubbed into the pork loins and bellies by hand. The loins were then cured for two to three weeks. The salting and drying of the meat results in a 30%

weight loss on the bacon. Unlike conventional bacon manufacturing, no water or phosphates are added to retain weight. To maintain the integrity of the meat, the Duchy bacon loins are not frozen, another practice used in conventional bacon production.

Distribution

In order to take the brand to the next level and to promote Duchy Originals on a larger scale, a focus on widening distribution is another key strategy employed by the business since 2000. Early distribution of the Duchy Originals biscuits started using select premium independent retailers. This developed a very strong core base of independents that still stock the brand today. As the brand developed, it was always the ambition to take Duchy Originals to a wider audience by using larger retailers, because to lead by example Duchy Originals always needed to be run as a serious commercial enterprise.

Broadening distribution of the range depended on a number of critical points:

- Communicating the Duchy Originals story to a wider audience and creating a high-profile organic brand presence in the supermarkets.
- Increasing the volume of Duchy Originals products being sold to generate a greater profit for the prince's charities.
- Leading by example and pushing for higher quality standards in supermarkets across categories and providing an organic alternative to conventional products.
- Exploring new product categories, especially in chilled product areas where organic products would be constrained by limited shelf life. For example, Duchy Originals ready meals have a limited ten-day shelf life and therefore require a relatively fast route to market that will generate a volume in sales that can justify the high cost of production.

Duchy Originals has established a strong relationship with Waitrose. This high-quality UK grocery retailer has a very similar ethos to the Duchy Originals brand. Today, Waitrose holds the widest range of Duchy Originals products and the retailer's online shopping service, OCADO, stocks over 75 Duchy products. Building on this, Duchy Originals is now also available on varying scales in all supermarkets.

Securing the right balance of distribution across major retailers and maintaining a strong independent base has been an important factor in the brand's success. Duchy Originals is still well represented in leading department stores such as Fortnum & Mason and John Lewis. It also has a strong base in organic retailers such as Fresh & Wild and Planet Organic as well as in small fine food or organic delicatessens. It still develops products that are exclusive to independent retailers, including a range of organic regional British cheeses, launched in 2005. The brand has extremely broad appeal from the premium market through to the health food and organic sectors.

Export distribution of the ambient range makes up around 6% of Duchy Originals sales. This is managed by partnerships with distributors in 15 countries worldwide, from Japan and Australia to the USA and Canada, and Europe. The products have been particularly successful in countries such as Germany where there is a strong eco-focus.

Awareness and promotion

Establishing Duchy Originals as a premium organic brand and raising awareness of the brand and product range is a key part of the company's marketing strategy. This has been supported by having extremely effective visual packaging, very strong brand and product stories and, ultimately, an excellent range of products.

Packaging

The packaging of Duchy Originals has created a very strong visual brand that gives it a unique identity. From the beginning, the packaging of each product has had to tell the story of the brand and communicate the quality and integrity of the products within. The visual look of the packaging is still based on the core concepts developed for the first Oaten biscuit. This has established a sound foundation for the brand that has since been replicated across other promotion material.

Telling the story

For Duchy Originals, telling a story on pack is essential to communicating why organic production is different and helping to reconnect people with the source of their food. The story creates the point of difference, illustrating how the brand has added value, as well as justifying the premium price point.

Packaging is one of the main vehicles for communicating the founding principles set out by the Prince of Wales and the overarching mission of the brand. All Duchy Originals packaging, however small, states: 'When HRH the Prince of Wales created Duchy Originals in 1990 it was because of his belief in the clear advantages of organic farming: the production of natural and healthy foods and sound husbandry, which helps to regenerate and protect the countryside and wildlife.' The Prince's Charities logo has also been an important feature of the packaging and is found on the front of each Duchy Originals product in a prominent place.

For meat and poultry products, the Duchy Originals packaging makes a strong point of describing the high animal welfare standards used as well as pointing out the source of the meat. For organic sausages, for instance, there is the statement, 'Duchy Originals organic sausages are made from organic free range pigs reared outdoors as part of the traditional cycle of mixed farming, helping to build and sustain the natural fertility of the soil. The pigs enjoy an organic diet, the highest welfare standards and a healthy, outdoor lifestyle on organic land. They have warm shelters with straw bedding and cooling mud baths in summer. Naturally reared and naturally fed.'

Each product package also contains a story, either of the special ingredients or the production methods that contribute to the unique character, flavour and integrity of the product. The Duchy Originals lemon curd has the statement, 'Made in Somerset to a traditional recipe with organic eggs, organic butter and organic Sicilian lemons'. In the case of ale, 'Duchy Originals Ale has been brewed in the traditional way, using a blend of the finest aroma hops and malt made from Plumage Archer barley harvested from selected organic farms in Britain including the Home Farm at Highgrove'. Ham packaging has the statement, 'Using a traditional recipe, the free range pork leg is cured for four days before being cooked and smoked over Cherrywood woodchips from the royal estates. This old-fashioned process produces succulent meaty rashers.'

Imagery

The brand decided to use etchings of pigs on both its sausage and bacon packaging to illustrate the environment in which they are raised. The picture association further reconnects the customer with the actual source of the bacon. At the time of the product launch, it was quite radical for a product such as bacon to include a picture of pig. Even today this imagery is much more accepted by an organic consumer who is happy to make the association rather than just seeing the finished product.

Clean ingredients listings

Duchy Originals also places a lot of importance on clean recipe lists for all products. Clean ingredients lists on organic products demonstrate the simplicity of production. For example, the Duchy Originals Damson preserve is made using only organic damsons and organic sugar. It is also important, where possible, to highlight on the pack the quantities of certain ingredients in addition to the standard ingredient listings. This shows that no corners have been cut in production, as is the case of Duchy Originals strawberry preserves where 60 grams of fruit is contained in every 100 grams. The Duchy Originals organic pork and rosemary sausages pack states that a 'minimum pork content of 75%' is contained each sausage.

Tactile packaging

Additional innovations have helped to endorse the individuality of the Duchy Originals and convey premium quality such as the use of parchment and paper style packaging to wrap the Duchy Originals bacon and sausages as opposed to plastic trays. The bacon packaging, when it was launched in 1999, set Duchy Originals products apart from its competitors creating a point of difference. The unique Duchy branded parchment gives the impression of the old style paper packaging used by local butchers.

Environmental packaging

It has been important to explore using recyclable packaging. Following customer feedback about the glass jars used in the Duchy Originals preserves, a new label was developed that could be peeled off the jar easily before it was washed to ensure that it could be reused. This type of label is now used across the preserve range. Recycling is an extremely important area for demonstrating environmental integrity. Duchy Originals is currently involved in a project with the Soil Association and Waste Reduction Action Programme (WRAP) to reduce the amount of packaging used in its biscuit packets.

Public relations

HRH the Prince of Wales has inevitably helped to attract media attention to Duchy Originals. The Duchy Home farm is an excellent platform for telling the Duchy Originals story and demonstrating the success that the prince has had in the value added marketing of products. The farm is now the model example of a successful, sustainable farm and business operation and attracts a lot of media interest.

PR activity aims to tell the brand story using product stories, communicating the provenance and quality of the ingredients as well as the processes of production. This has been achieved through unique stories, such as the sourcing of honey from the Balmoral estate in Scotland. The products always consistently rate well with the food media and are positioned as premium quality.

The Duchy Originals website is becoming an important promotional tool (www. duchyoriginals.com). This has been the most comprehensive mechanism for communicating the whole Duchy Originals story. Each product is included on the website with a description, image, ingredients and nutritional information. The site also includes detailed information on the founding of the brand, the prince's philosophies and the principles of organic and sustainable production as well as the stories of our partnership with manufacturers, such as Walkers, to produce the range.

Advertising and promotion

It was always anticipated that once it was commercially viable, the possibility of advertising the brand would be explored. The first Duchy advertising campaign took place in 2005, targeted at raising awareness of the Duchy Originals brand and stimulating trial. The executions featured biscuits, bread, ham and sausages and were launched in the lead up to Christmas 2005 on taxi sides, in leading food and lifestyle publications and in poster shells in key locations.

The philosophy of the Duchy brand was embodied in the campaign strapline; 'Duchy: The way life should be'. Individual executions used the provenance of premium ingredients as key selling points. For instance, Duchy Originals Sunflower & Honey Bread is promoted using the copy, 'Crafted with organic sunflower seeds and honey. Our bread is anything but run of the mill.' Another execution, for Duchy Selections free range ham, describes the use of special ingredients such as clementine marmalade and heather honey.

In 2006 the brand also embarked on its first sampling and exhibition campaign, recognising the value in sampling activity to stimulate trial and raise brand awareness. The challenge with marketing a premium product is that the price point can deter some consumers. Once people taste a Duchy Originals product they are motivated to purchase. Sampling the products at a small number of key events throughout the UK, including the organic food festival in Bristol, will be the focus of the first year of the sampling activity to test the water.

Consumer insights

Ultimately, consumer research conducted in 2004 suggests that the main reason people buy the Duchy Originals range is because of its premium quality and taste. The range appeals to a much wider audience than solely organic consumers. The organic status is, however, an underlying endorsement of the premium quality of a product.

Organic and environmental considerations are a secondary motivation for most consumers buying Duchy Originals products. Consumer feedback suggests, however, that the importance of organic rises on their purchasing agenda mainly because of food safety concerns. This is especially important when considering fresh produce purchases, such as meat, where there is a greater concern from people about the use of antibiotics, hormones and animal welfare standards.

There was a very high awareness amongst consumers surveyed about the link of Duchy Originals to HRH the Prince of Wales. The feedback received, however, suggested that though this might prompt some customers to try the range, it was premium product quality that prompted repeat purchases. While the charity is not a primary motivator to purchase the products, it enables consumers to feel that they are giving something back to the community.

Future growth

The ambition for the Duchy Originals brand for the period 2006–2009 is to increase its profit donation to the prince's charities to over £2m per year. To achieve this,

Duchy Originals will continue to focus on enhancing its product range, increasing the distribution channels and exploring new distribution opportunities, and enhancing its promotional activity through further advertising and sampling.

The brand also embarked on an innovative development in June 2006 when it launched its first Duchy Originals bakery in Launceston, Cornwall. The launch signalled the first instance in which Duchy Originals had a direct hand in the development and management of its own production facility. Duchy Originals Foods Limited is a wholly owned subsidiary of Duchy Originals and produces a range of organic pastry products.

The decision to open the bakery was threefold: the brand was looking to produce a range of pasties and pastry products organically, to the quality standard required of a Duchy Originals product, and at the quantity that could support its main retail customers while maintaining product integrity. The brand took a decision to establish the bespoke organic bakery after struggling to find an existing producer that could meet all of these objectives.

The Duchy Originals bakery in Cornwall will work alongside other Duchy Originals manufacturing partners. A number of other Duchy Originals products will be incorporated into the range including a lemon tart made with Duchy Originals lemon curd and a cheese and bacon flan with Duchy Originals bacon.

Measuring brand success

The association of the Prince of Wales has been invaluable in building the profile of the Duchy Originals brand. Having such a strong spokesperson to support the brand with such unrelenting passion for promoting organic and sustainable high-quality food production is extremely powerful. This poses the question, however, would the Duchy Originals brand model be a success if it wasn't linked to the Prince of Wales? The answer is, yes. For Duchy Originals, maintaining the highest product quality and adhering to the founding principles of the brand is fundamental to the brand's success.

The whole idea behind the Duchy Originals brand was to lead by example and demonstrate that it could be done. It is evident from the success of other brands that have a similar ethos and which have developed in parallel to Duchy Originals, such as Green & Black's, Rachel's Organic, Yeo Valley and a wealth of other premium organic producers, that it can be done.

Conclusion

The link to HRH the Prince of Wales ensures that Duchy Originals products will always have an underlying integrity. Yet, at the end of the day, according to Chief Executive Belinda Gooding 'if they are not fantastic products, they won't sell and shops certainly wouldn't continue to stock them.' The critics seem to agree and this is perhaps the greatest measure of success. Over a decade after the launch of the Oaten biscuit the product was awarded a silver award in *The Daily Telegraph* Taste of Britain Awards in the best UK food category 2005. Duchy Originals has established its place as not only a pioneering brand in the organic food movement but also as an icon for UK food and the renaissance of a quality British food culture that continues to flourish. This is something the humble Oaten biscuit, founded with a very serious mission, can be very proud of.

References

1 Smith, D. (1992) The biscuit that Charles baked. *Taste Magazine*, December, p. 16.
2 Blythman, J. (1993) Prince's Originals take the biscuit. *The Independent*, 8 May, p. 32.
3 Wright, S. and McCrea, D. (Eds) (2000) *Handbook of Organic Food Processing and Production*. Blackwell Publishing, Oxford.
4 (2006) Soup category review. *The Grocer*, 17 December, p. 119.

Chapter 9

Case History: Sainsbury's SO Organic

Ruth Bailey
Senior Brand Manager, Organics, Sainsbury's Supermarkets plc

Background

The Sainsbury's story begins in 1869 when John James and Mary Ann Sainsbury opened their first small dairy shop at 173 Drury Lane, London (see Figure 9.1). Since then Sainsbury's have evolved into one of the UK's leading multiple food retailers with sales of £17317m (including VAT, financial year end 2005–2006), an estate of 752 stores (455 supermarkets and 297 convenience stores) and 153000 employees.[1]

Sainsbury's is today as passionate about what it sells as when there was only one store. It is therefore fitting that Sainsbury's were the first retailer to introduce own-brand organic and fair trade food to the UK population and continue to be at the forefront of industry leading developments today.

Sainsbury's organic journey began in 1986 when Sainsbury's food technologist Robert Duxbury visited a small vegetable farmer in west Wales. It was from this farm that Sainsbury's bought its first organic vegetables. The farmer was Patrick Holden. Today he is the Director of the Soil Association (the UK's leading organic certification body). Over the past 20 years, Sainsbury's has grown its organic range from this one regional carrot to over 350 own-brand products in over 600 stores across the whole country.

The successful history and evolution of Sainsbury's organic range is evident in the continuous sales growth that has been achieved and the fact that Sainsbury's organic market share has consistently outperformed the corporate market share. By mid 2006, Sainsbury's total organic market share was 30.9% (ACNeilsen 52 week data to 20 May 2006) compared to a corporate market share of 14.7%.[1]

Impressive sales and market share performances are not the only evidence of Sainsbury's market-leading position, both organic-specific and wider, general industry bodies have recognised and acknowledged Sainsbury's strength, above other retailers, within the market. Over the last decade, Sainsbury's have been awarded with a Quality Food and Drinks Awards, Gold award for Sainsbury's Organic Somerset brie (2003) and have received the Soil Association's Supermarket of the Year award

Figure 9.1   The first Sainsbury's store: Drury Lane, London.

for a number of years (Figure 9.2). Although Sainsbury's have experienced great success with their organic range, they have never stood still and 2005 was to be a year of great change that would significantly improve the offer to drive even greater success.

The need for change

For the whole of the Sainsbury's company, 2004 was a year of great change. A new management team were in place with ambitious but clear plans to strengthen Sainsbury's own-brand product range. An exemplary own-brand product range is fundamental in influencing the perception of any retailer. Own-brand products are a direct reflection of the company; they communicate what the company is focused upon and are a powerful tool in differentiating one retailer from another. For customers, they provide a tangible way to understand the differences between retailers, therefore allowing them to make an informed choice as to where to shop.

Within Sainsbury's own-brand offering, there are a handful of sub-brands. The role of the sub-brands is to clearly segment the market, allowing the customer to easily navigate product ranges to find the products that most fit their need or desire. For example, Sainsbury's offer a 'Taste the difference' range that delivers the ultimate in everyday food – the best tasting food, using the most authentic ingredients and traditional cooking techniques. To complement this, a value range called 'basics' is also provided. This range offers competitive, lower-priced everyday essentials without compromising on Sainsbury's core quality standards.

Sainsbury's organic range had done a successful job of differentiating itself against the competition and securing a level of customers higher than it's fair share. Organic food and drink sales at Sainsbury's account for a higher percentage of total food sales than within most other retailers (TNS Superpanel 52 week data to 23 April 2006: Sainsbury's 1.29%; Tesco 0.77%; Marks & Spencer 1.23%; Asda 0.39%; Waitrose 3.22%). Sainsbury's customers, in general, actively seek higher quality food, so

Figure 9.2   The Soil Association Supermarket of the year award.

organics has always been disproportionately important to Sainsbury's. Sainsbury's had succeeded in further differentiating their organic offer by establishing a strong brand identity with easily recognisable packaging and had earned credibility by offering customers more organic choice than they could get in other retailers.

But by 2005, the landscape had shifted substantially. Consumer interest had risen consistently over the past 20 years as more and more customers were appreciating the benefits of organic food. As a result the market was growing at over 20% year-on-year (ACNeilson 52 week data to 20 May 2006: total organic market growing at 21.4%). Consequently organics was no longer a niche market and the resulting sales opportunity was attracting the other multiple retailers who were driving forward with their own organic ranges.

The existing Sainsbury's organic packaging had been on the shelves for many years. It had not evolved at the same pace as the market had (or other Sainsbury's sub-brands such as 'Taste the difference') and as a consequence it no longer reflected our customers motivations for purchasing organic food and drink, nor was it enticing new customers into the market.

It was the combination of these three factors – a corporate strategy to use sub-brands to differentiate the Sainsbury's offer, a market experiencing phenomenal growth that was appealing to a broader range of our competitors and an existing organics range that needed a refresh – that lead to the relaunch of Sainsbury's organic range in September 2005 as 'Sainsbury's SO organic'.

Understanding our customers

This process began in the autumn of 2004 with a deep-dive customer research project. The aim of this work was to fully understand both the organic market and who the organic customer was. Questions to be answered included:

- Why was the organic market growing so fast?
- Where was the majority of the growth coming from?
- Who was buying organic food – were there more females than males buying organics, were they younger or older?
- Why were they buying organic food – were the reasons different for different groups?
- Had the reason to purchase changed over the last few years?
- What did customers think of the current Sainsbury's organic offering – what was good, what was bad?
- According to customers, how could Sainsbury's improve their organic offering?

Answering these questions was achieved through a combination of qualitative and quantitative sources. Quantitative sales and market share data of other retailers and different customer groups was combined with qualitative panel surveys that asked the same attitudinal statements (e.g. Do you think that organic tastes better?) to many different customers.

More exploratory questions (e.g. How can Sainsbury's make their organic range better?) were asked through smaller samples of focus groups. These groups also helped with the content and creation of mood boards. These boards were collages of pictures, photographs and other images that represented and reflected these customers' understanding of what organic food is and the benefits of it, plus their personal motivations to purchase it.

We had a clear historical picture of how the organic customer had evolved over the last couple of decades but the research showed how significantly attitudes, perceptions and therefore the type of consumer buying organics had changed over the most recent five years.

In the 1970s and 1980s organics had stereotypically attracted a relatively specific type of consumer – typically these were regarded as new-age or 'hippies' that had strong ethics and morals regarding food production and the environment. As a result, organics became characterised as 'do-gooding' with connotations of sandal wearing and tree-hugging hippies.

By the 1990s, while still a young market, organics had attracted a second, larger group of consumers. These were an older (35+), well-educated and relatively affluent sector of society (socio-economic class ABC1). This group (commonly referred to within the industry as 'dark greens' due to their strong commitment to organics) shared the 'hippies' environmental motivations for purchasing organics. They were likely to have recently started their own families, which triggered concerns about protecting the world for the next and subsequent generations, but more influentially, they were the first consumer group to purchase organics for another reason – a reason that was to change the face of organics.

By the mid 1990s, the general public were paying more attention to health and well-being. The press were printing sensationalist articles about rises in cancer rates, links with illnesses, childhood obesity and intolerances and our modern lifestyles, e.g. pollution, nutrition, smoking, etc. Consumers viewed organics as a positive choice that they could make to contribute to both their own and their families' health and well-being.

The research conducted by Sainsbury's in 2004 proved how influential health and wellness had become as a reason to buy organics. There was little evidence of the original environmental campaigning hippies. The 'dark greens' (now aged 45+ years) dominated the organic market (accounting for 50% of both the spend on and consumption of organic food and drink) but their purchases were established and therefore growth rates were relatively static.

Although accounting for the majority of the organic market, these customers only accounted for a small percentage of the total population and therefore in isolation couldn't have driven the levels of growth that the organic market was experiencing by the new millennium. The continuous growth could only be accounted for by a mass of new customers buying into the market. It is the needs and motivations of these new customer segments that stimulated and underpinned the launch of Sainsbury's SO organic.

'Young families' is one of these new groups of consumers. The life-changing event of the birth of a new family member, in particular the first child, has always played a pivotal role in changing shopping habits and parents regard going organic as a positive way to give their baby the best start in life.

For these customers, their motivation for purchasing organics mirrors the trend that emerged from the 'dark greens'. In a modern society plagued with headline grabbing stories of numerous health issues that we face on a daily basis – many of which we have little control over – parents feel a strong emotional drive to protect and nurture their immediate family. They view organics as a positive action and one that they can actively manage.

Most recent has been the emergence of the 'pre-family' customer segment. Although these customers currently spend the least on organic food, they are the group that are showing the greatest growth. Generally young, free and single, these are a group who buy into food as fashion – what they put into their shopping basket or on the dinner party table is an overt statement of who they are or how they want to be perceived. The upsurge in celebrity endorsement (both by celebrity chefs and Hollywood stars) of organic food has driven their interest in this food sector.

As a result these customers are the least knowledgeable about organics and as such their driver for purchase is more divided than the other customer segments. 'Pre-families' purchase partly for health reasons but for the majority of these customers, it is their belief that organic food tastes better than conventional that drives them to purchase. A percentage of the other customer segment groups cite 'taste' as a driver for purchase, so this is not a factor unique only to the 'pre-families' but it is displayed more prominently within this group than within any other

The research conducted by Sainsbury's demonstrates that the original, environmental drivers have been superseded. The dominant driver for purchase is now concerns regarding health and well-being. Although not the primary driver for the majority of customers, the belief that organic food tastes better is very influential, especially for the newest organic customers. That said, environmental factors remain very much a foundation of organics for many – old and young, established and new. To be a market leader, a retailer must continue to appreciate, respect and respond to this customer need.

Before progressing further it is critical to cast a caveat over the information above by clearly stating that although health and well-being is the strongest motivation to purchase, price still remains the greatest barrier to purchase, and for some people the deterrent of price overrides the motivation of health.

This is a marketing conundrum that the organic movement has constantly struggled with and will continue to do so for the near future. In pure economic terms, the processes and procedures involved in organic food production do cost more, so the end product also costs more. In addition, current supplies of organic food do not outweigh demand to the same extent that they do within conventional food markets – this also pushes up costs.

From a marketing perspective it may seem logical to lower the price of organic food in order to attract more customers. Any retailer considering this tactic needs to do so carefully – the popularity of organic food is based on strong reassurances regarding how that product is produced (i.e. the product will be certified as organic by an approved certification body). Should retailers embark on a mass price-cutting exercise, customers may start to question the quality and authenticity of the product.

Given that price wars could be so destructive, the key to sustained growth is to deliver products that exceed customer's expectations on taste and to educate customers about the benefits of organic food and the additional costs involved in producing it.

Building the Sainsbury's organic proposition

Delivering a successful brand relies on listening to customers to meet and exceed their needs. Listening and learning from customers will also provide insights into emerging trends. Retailers can then anticipate and develop solutions to these impending needs, satisfying customers before they realise it, thus securing longer term loyalty.

The customer deep-dive was crucial in ensuring that we understood organic consumers in as much detail as possible. Once armed with all this knowledge and understanding, Sainsbury's could then create plans that were sure to exceed all customer demands and make Sainsbury's the customer's favoured organic retailer.

Next in the process was deciding what we wanted the Sainsbury's organic brand to stand for. What did we want people to think and say about it? From the beginning it was apparent that health and well-being would play a central part, but as one of the UK's supermarket retailers what more could we do to make it easier for our customers to go organic everyday?

Aside from the motivation to purchase, the research (and phenomenal sales growth) highlighted that the profile of the organic customer was changing. There was a core group of committed, older, more affluent 'dark greens'; young families wanting to protect their offspring and the lifestyle-driven 'food as fashion' brigade. In essence, organics was no longer a niche market; it had become mainstream and had obvious universal appeal to be able to attract such diverse consumer groups.

Therefore, Sainsbury's proposition consisted of three clear elements:

(1) Keep the motivational driver of health and well-being at the core of everything we do. The word 'health' in itself means a myriad of different things to different people. Therefore for the Sainsbury's organic proposition to provide clear, actionable direction for the business it was vital to clarify what this was and wasn't. We understood that customers believe that the fresher food is, the more goodness it contains, and that the production methods that characterise organics deliver more natural products than modern-day mass food manufacture. For Sainsbury's, organic health is about providing the most natural and freshest

food that we can. It is not about applying the science of health, e.g. fortifying organic juice with additional vitamins and minerals.

(2) Make organics accessible to all. Organics is no longer a niche market dominated by one type of customer. The proliferation of different customer groups provides a lucrative sales opportunity but at the same time it adds complexity. The different customer groups have varying degrees of commitment and differing purchasing patterns. For example, young families will buy more baby food than anyone else, understandably pre-families regularly purchase chilled pizzas and fresh pasta and the over-45s purchase more fresh fish than any other customer group on a like-for-like basis. The Sainsbury's organic brand is committed to offering solutions for all of these customer groupings.

(3) Remember that price is the greatest barrier. Relatively speaking it's easy to offer the greatest choice of the freshest food available, but given that price is the biggest barrier to purchase, this would be pointless if the retail charged became unreasonable and unattractive to customers. In addition, with an aim to make organics accessible to all, we consistently face the challenge of offering compelling value (in order to attract current non-adopters) while remaining conscious of not sparking a price war or devaluing the whole organic market.

A proposition is critical to providing the organic brand with a purpose and personality but in essence it is a statement – in order to make it a reality, a supporting set of governing rules are required to ensure that all products developed are produced to consistent standards and marketed to customers in a uniform manner.

The organic brand standards support our drive for delivering freshness through a declaration to favour sourcing from the within the UK, which limits the amount of time (and food miles) that our produce takes to get from the field to our consumer's plates. In addition, Sainsbury's are proud to have taken this to an even higher level by guaranteeing that 100% of our chicken, beef, pork and milk are UK sourced (Figure 9.3).

Figure 9.3   Sainsbury's SO organic is strongly committed to sourcing from within the UK.

Our brand standards also empathetically address the environmental elements of organics that still remain hugely important to certain customers today. Sainsbury's actively encourages the use of recyclable and/or compostable packaging materials for products within the organic brand and are proud to be at the forefront of pioneering material development such as the use of a compostable material made from non-GM maize, which we are currently trialling on Sainsbury's SO organic apple and potato bags.

Our rule-set ensures that we are making organics accessible by stipulating that a minimum of 100 products are offered in all of our 450+ supermarkets. This ensures that all Sainsbury's customers, wherever they shop, can confidently do so knowing that they will find an organic version of their entire staple shopping basket items.

The price conundrum is addressed through the rule-set. This provides a clear framework as to how to set retail prices to ensure that across the company they are relative, reflect the true costs of organic production and that value is offered in comparison to third-party organic brands and the competition's own-brand organic ranges.

Once the Sainsbury's organic brand had a personality (through the proposition) and strong principles and direction for what it is and isn't (the governing rules), it then needed an identity – how would our Sainsbury's customers be able to identify the range and what would it look like?

The development of the Sainsbury's SO organic name and design

Sainsbury's existing packaging design had worked hard for the brand over a number of years but due to the fast paced change within the organic industry and the pace of change of other Sainsbury's sub-brands, the design no longer achieved stand-out on shelf nor did it reflect the modern, contemporary or universal appeal that today's organic market demanded.

There were three distinct elements to the new organic design – the brand name, the graphic packaging design and the photography (Figure 9.4).

The name chosen, Sainsbury's SO organic was the development of many ideas. It imaginatively combines the 'S' of Sainsbury's with the 'O' of organics to create 'SO', but it is much more than a clever play on words. The word 'SO' symbolises Sainsbury's commitment to organic and is a statement of intent that not all organic products are the same and that Sainsbury's strives to take the higher ground and provide customers with the highest organic quality at fair prices.

For the packaging design we deliberately chose a green palette reflecting the essence of the organic market, the earth and the grass. But, a dark green, rather than primary green was used for a number of reasons:

Figure 9.4   The Sainsbury's organic packaging after the relaunch.

- The darker green creates a premium feel and as such represents the quality aspects of the brand
- The darker green provides striking stand-out on shelf, making it easy for our customers to locate the organic products
- The darker green is a versatile colour that works well across many different pieces of customer facing communication such as in-store point of sale.

A vignette (gradual lightening of the colour) was used across all product packaging. This effect adds to the premium cues as well as creating an earthy feel across the design.

The photography is consistent across all products and is a single shot of the source of the product. For most products this is the plant that the product is produced from. This photography is an icon of freshness and provenance as it shows where the product came from in its most natural source.

The relaunch

With the strategy set and design confirmed, the relaunch event planning began in earnest. The relaunch was to land in stores in September 2005 – a natural calendar event that has historically celebrated the seasonal change in food production, symbolised by the harvest festival. The customer research and insights used to develop the personality and identity of the brand were a natural source of inspiration for the theme of the in-store launch event.

The first prong of the event was to ensure that we were making organic accessible to all. In particular this meant identifying product gaps within the range and developing the relevant products. In total we launched over 100 new lines between the beginning of August and the relaunch in September.

An interesting case study was the development of the Sainsbury's SO organic wine range (Figure 9.5). Historically, wine was not a category that attracted much organic attention or stocking of product. If considered within the context of the newer organic customer groups such as the 'pre-families', however, increasing customer awareness of the implications of sulphates in wine and the established universal appeal of organics, an organic wine range seems fitting. The fact that Sainsbury's had the impetus to develop such a range within the organic umbrella is testimony in itself of the development and evolution of the market and Sainsbury's commitment to driving the organic market forward and continuously meeting customer needs.

Developing new products was not enough to ensure that we were making organics accessible to all. The complete list of organic lines being sold were analysed and from this a commitment was made to ensure that a specific selection of over 100 lines would be available in all of our supermarkets.

The second prong of the event would highlight to customers our commitment to sourcing British, thereby providing a tangible choice for our customers to buy and

Figure 9.5 Sainsbury's new wine range. Just one of over 100 new products added to the range in September 2005.

enjoy the 'freshness' of organic food. Finally, all activity communicated was supported and bolstered with the credibility of a value message, bravely tackling the inherent barrier to purchase. To deliver this, Sainsbury's itself (not its supplier base) invested over £10m in permanently lowering the prices of over 100 staple shopping basket items including bread and carrots, foods that are real entry points for our customers, which has helped make it easy for them to go organic.

Merchandising and signage

The store layout and merchandising of the brand was revised. The market data confirms that nearly 80% of the UK population purchase at least one organic item every year (TNS Worldpanel, 52 week data to 23 April 2006: total organic market penetration: 78.7%). Very few of these purchases, however, are from loyal and committed organic shoppers – 68% of organic spend comes from 12% of households who purchase organic (TNS Worldpanel, 52 week data to 23 April 2006), indicating that there is real opportunity to open up organics to more customers, more often.

Accordingly it was decided that customers would be more likely to purchase an organic product were they merchandised alongside their non-organic equivalents. In addition, the customer insights had revealed that customers found organic items hard to find and others only made the decision to purchase an organic product once they were at the fixture – they were not pre-determined decisions.

An organic dedicated aisle may initially seem more attractive as it provides customers with a 'one-stop-shop' but this layout can be polarising. For example, if you don't have dogs and cats you don't go down the pet food aisle. To actively choose to go down the organic aisle requires the customer to declare themselves as organic consumers. There are a far greater number of 'dabbling' organic shoppers than committed organic consumers, so the greater commercial opportunity is to satisfy the dabblers and make it easier for customers to convert more of their shop to organic.

The exception to the rule is within areas that we know customers make a pre-determined decision to purchase organic. These are categories with a critical mass of lines such as produce (fruit and vegetables), meat and parts of dairy (e.g. yogurts). These three areas deliver a disproportionate percentage of organic sales because they are categories where the organic concept of field to shelf and the benefits of organics are easiest to understand and as such, are the first categories that new organic shoppers will buy into. This makes these categories destination areas within the supermarket, thereby justifying the decision to merchandise together and warranting a point of sale advertising toolkit to add theatre and highlight to customers (Figure 9.6).

The point of sale advertising toolkit for these areas included permanent header signage above the section to ensure that customers could locate these key, destination categories. Hard-wearing plastic dividers were used to hang vertically at each end of the organic blocks. For areas that occupied only a couple of shelves, 'bookends'

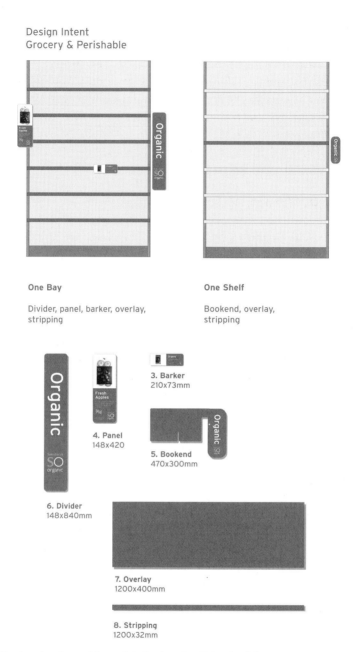

Figure 9.6   Design drawings of the point of sale advertising toolkit.

(a smaller version of banners) were used. Paper stripping with reproduction of the Sainsbury's SO organic logo was produced for the edges of the shelf to assist customers with finding the organic versions of product and provide a symbolic visual aid to assist with recognition.

In summary a new, fresher design with more impact had been designed and was to be unveiled across the whole own-brand range, including 100 new products. Over 100 lines were available in all stores, £10m had been invested in lowering prices on everyday essentials, the commitment to British was being driven forward and the merchandising and corresponding permanent point of sale had been constructed.

Once all the content was confirmed, it was used to generate the customer facing communications both in-store and externally for the launch event. These communications encompassed all three prongs of the event principles.

Customer communication

The above the line national communications efficiently combined two messages bringing them together as one clear customer execution: quality organic foods at lower prices. Figure 9.7 shows the creative expression of both a quality focused British sourcing message along with a value stream communicating lower prices. There were four executions of this combined message: carrots, bread, potatoes and chicken.

In addition to the national press, a local press campaign ran at the same time in London and the South East, regions chosen because of their heavily biased organic customer base. The regional executions took the same theme as national (a combined British and lower prices message) but a slightly different creative execution was created.

This execution was predominately used in high-traffic commuter areas such as the London Underground and London transport bus shelters. These sites are viewed by key organic customer segments such as young families and pre-families, who have higher than average incomes and lifestyle interests that naturally fit with organics.

Aside from advertising, co-sponsorship (with the Soil Association) of a supplement within the Saturday edition of *The Independent* newspaper provided an editorial platform, reaching thousands of potential customers that enabled us communicate richer content such as the reasons to go organic and present articles that reflected the shift to organic as a modern and positive choice.

The co-production with The Soil Association gave credibility to Sainsbury's position as a retailer serious about organics. This is a result of the prestigious position that the Soil Association holds in the eyes of consumers as the UK's leading certifying body.

The above the line messages of British and lower prices were created, as with any above the line media, to drive awareness. These headlining messages were reiterated within the below the line media to create a stronger drumbeat and resonance with customers. In contrast, the strategic objective of the below the line media was to drive sales and conversion.

Sainsbury's magazine was used to showcase the breadth of range that Sainsbury's SO organic offered – including some of the iconic new products with recipe sug-

Figure 9.7   National Press advertisement. Courtesy of Sainsbury's and AMV. Photograph by Jason Lowe.

gestions – and highlighted our strong brand standards by showcasing some of the suppliers that Sainsbury's are proudest to work with, such as farmer Henry Gent and his Devonshire gold organic chickens.

To add theatre within stores during the event, an in-store activation agency was used to deliver a customer sampling programme (Figure 9.8). This activity gave

Figure 9.8 The customer sampling programme.

customers currently not buying organics the chance to experience the 'taste' difference that many organic customers extol and for committed organic shoppers it highlighted some new products that they may not have seen before. Established brands such as Yeo Valley and Twinings were sampled alongside Sainsbury's SO products with over 120 000 samples given to customers in over 30 stores. Products sampled ranged from milk to wine, creating a fun and interactive way to launch the Sainsbury's SO organic brand.

Colleague engagement and belief

As one of the UK's leading supermarket chains, Sainsbury's has 153 000 colleagues in 752 stores[1] who are the 'front-line' of communication to our customers. Therefore, the success of any in-store event or product launch is highly dependent on colleague engagement and developing colleague belief. In the case of Sainsbury's SO organic we wanted colleagues to understand and deliver our organic vision in-store, merchandising products in the right places and enthusiastically highlighting the range to customers. To achieve this we needed to drive colleagues to be passionate about organics through providing them with an understanding and of our new organic range and its brand identity.

This communication was all contained within a single briefing pack that was sent to stores, which store managers disseminated among their teams. This ensured clarity of message and provided one definitive source to refer back to. Stores were provided with an incentive scheme that rewarded the store within each geographical region that delivered the greatest percentage uplift on organic food and drink over the launch event in September. In addition, we showcased the new range by offering organic options in staff canteens during the event and by holding a 'farmers' market' at the central office, which gave our organic suppliers the chance to talk about their products to our colleagues.

The results

Results following the relaunch have been impressive. In terms of sales, the Sainsbury's SO organic brand has consistently delivered year-on-year sales uplifts of 20% with the total organic offer (third-party brands included) delivering double digit year-on-year growth (Source: Sainsbury's trading finance department). Since the relaunch, growth rates have been ahead of the rest of the market.

A major highlight of 2005 was the performance at Christmas. We actively increased the range of lines offered to ensure that all customers could enjoy a completely organic Christmas, supported with an in-store leaflet highlighting the whole range. Customers responded positively and our committed organic customers appreciated the breadth of range offered, while there was evidence of non-organic

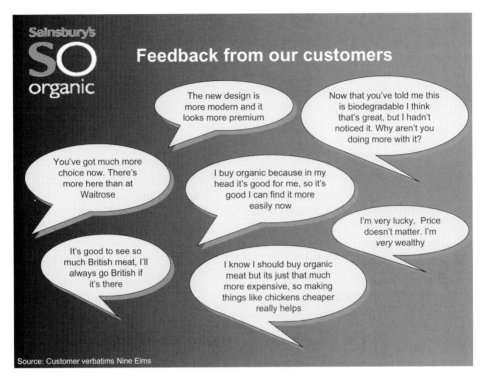

Figure 9.9   Feedback from SO Organic customers.

customers viewing the Sainsbury's SO organic range as a quality cue and thus a trade-up to treat themselves, their friends and families with at this seasonal time of year. Organic sales for Christmas 2005 were 40% higher than the previous Christmas and the highest levels achieved across the whole year.

Ultimately, as a retailer, the most gratifying feedback comes from our customers. While the event was live in stores, accompanied customer shops took place to assess levels of awareness and response to both the event and the new Sainsbury's SO organic range. The feedback given unanimously confirmed that customers regard the new-look packaging as a positive improvement and a sophisticated and contemporary interpretation of modern-day organics that has greater stand-out and is therefore easier to find in store.

Figure 9.9 gives customer comments including, 'You've got much more choice now. There's more here than at Waitrose', 'The new design is more modern and it looks more premium', 'I know I should buy organic meat but it's just that much more expensive, so making things like chickens cheaper really helps', 'It's good to see so much British meat, I'll always go British if it's there.'

Another important measure of success is the performance of Sainsbury's organic range compared to the competition. In terms of market share, Sainsbury's own-brand organic range (Sainsbury's SO organic) saw market share percentage increase from a level of 26.8% in August 2005 before the relaunch, to a high of 32.2% during Christmas 2005 post the relaunch, which outstrips Sainsbury's total market share (Source: TNS Superpanel).

When compared to Sainsbury's total market share, the appeal of the organic range is dramatically highlighted. After two years of holding the number two market share position (behind Tesco), Sainsbury's SO organic took own-brand market leadership for the 16 weeks to 26 March 2006 (TNS Superpanel, average weekly market share of 31.9%) making us the UK's favoured organic retailer. To complement the market share improvements, market data shows that Waitrose customers have actively been switching their purchases to Sainsbury's organic ranges.

The industry has also been praising the SO organic brand since its arrival. The new brand design won the 'Design week' award 2006 for 'Own-brand packaging' and the Sainsbury's SO organic brand has been a finalist in both *The Grocer* Gold awards and *The Retail Week* awards.

Future challenges

For both the organic market and specifically for Sainsbury's organics, 2005 was a momentous year. Subsequently the market has continued to experience strong growth and the competition has responded accordingly, however, thereby keeping the pressure upon Sainsbury's to constantly strive forward, consistently delivering for our customers both in terms of quality, choice and price. The market looks set to continue with double-digit growth for the foreseeable future, securing its place at the centre of all retailers' interests and thoughts.

An immediate challenge that faces all retailers within the organic market is the availability of product. The sharp increase in customer demand has been relatively recent; conversely the transitional period for a farmer or producer to convert their land and processes to fully organic can take many years. There is currently a time lag between increasing customer demand and farms converting, resulting in an immediate availability issue. The challenge then falls to the marketing teams to keep customers engaged, passionate and committed to organics while that time lag is closed.

Sainsbury's has been particularly active within this arena and in April 2006 launched a milk product from farms in the process of converting to organic. 'Farm Promise' milk provides the farmer with a Sainsbury's contract that commits to financially support the farmer through the conversion process and also a commitment to take the farmer's milk once it is organic. This initiative gives the farmer the confidence and financial security to embark upon the conversion process. Sainsbury's have also written contracts with organic meat suppliers, providing those farms with the confidence to convert and also developing long-term and trusting relationships between farmer and Sainsbury's.

The organic market is no longer a niche market, with many differing customer profiles and groups now organic customers. Those marketing organic products to these customers need to ensure (as with any other market) that they are tailoring the content of the marketing messages to the needs of the differing customer groups.

The polar differences between the established 'dark greens' versus the pre-families need to be respected. 'Dark greens' are highly knowledgeable and informed about what organic is and isn't and have strong reasons for purchasing. Conversely, pre-families buying organics as a lifestyle choice, have low levels of understanding of what being organic means and as such may have less appreciation of the price premium attached to organic food versus conventional food. The content and tone of voice used to communicate to these differing groups needs to be carefully considered.

Longer-term challenges will include the positioning of the organic sub-brand. As organics becomes more mainstream and a part of customers' regular shopping repertoire, customers will not want to make trade-offs or compromises.

For example, customers may want a premium, indulgent and quality Christmas pudding such as one laced with cognac but they may also want this to be organic. Mums may want crisps that are developed specifically for kids, but they may also want these to be organic. For a supermarket, this raises a challenge as to whether organics should continue as a separate sub-brand or whether it should be combined with other sub-brands that have close connections with organics such as premium or kids' brands.

Finally, the large supermarket retailers will have an additional job keeping customers purchasing organic products within their stores. There is evidence to show that independent organic retailers and retailing channels such as the small sole-trader stores, farmers markets and shops are rapidly on the increase and are prospering. In 2004, these channels took trade from the supermarket chains, reducing the percentage of the total organic market that was accounted for by supermarkets. For some customers these independent channels offer a more personal shopping experience that they don't believe that they get from a supermarket. The supermarkets' challenge in response is to retain the convenience of one-stop-shopping but to enhance the customer's visit, to provide an experience and to develop ranges that come from various sources, promoting British while also supporting smaller producers and embracing and celebrating local and regional differences.

Conclusion

The organic market is the most exciting food sector within the UK at present. No other sector is experiencing such levels of growth, media coverage or customer interest. For those involved in the industry it provides a dynamic environment in which to work and contribute. The immediate challenges such as keeping supply at pace with demand sit alongside longer-term visionary thinking, creating plans that will continue attracting new customers, while retaining the committed consumers and converting the dabblers into loyalists. The organic sector is growing so quickly that today's long-term, visionary thinking will be tomorrow's activity and events.

Reference

1   J Sainsbury plc (2006) *Annual Report and Financial Statements 2006*. www.j-sainsbury. co.uk/ar06/fullfinancials/financialstatements.shtml

Chapter 10

Organic and Fair Trade Marketing in Germany

Hubert Rottner

History

In Germany, the first organic food stores were established in 1975. Today, Germany's annual turnover of organic products amounts to €4bn. This compares with €11bn in the USA and €23bn worldwide. Germany makes up 30% of the total organic market in Europe.[1]

In the early days there was little money for marketing of organic food and it was often felt that, anyway, marketing was not 'cool' and was rather a waste of money and resources. The most effective method of marketing was by word of mouth. Still to this day most companies in Germany are responsible for their own promotion.

The first big overall attempt to promote organic produce was between 2001 and 2005, when the red–green coalition government and its Minister of Agriculture and Consumer Protection, Renate Kühnast (www.bmelv.de), took many steps to support organic agriculture, as well as to promote organic products and lifestyles. One effective step was the implementation of a unifying national organic label, the so-called 'Biosiegel' (Organic Seal) based on the current European standards (April 2006).

In Germany this label is now used by 552 manufactures for more than 31 400 products (for more information see: www.bio-siegel.de). This development proved a most effective marketing action as it simplified the former confusion of organic labels in Germany. Now, as part of the Biosiegel process, grower association labels are used as premium labels because their standards are higher than the European legislation for organic products. Full details of the programme, which is published as an internet database, can be found at www.oekolandbau.de. For organic scientific support see www.bundesprogramm-oekolandbau.de.

This government programme had a budget of €80m spread over four years and should be continued by the Big Coalition with Horst Seehofer as the responsible Minister.

The eight organic grower associations

In 2006, 16 791 farms with 800 000 hectares of organically farmed agricultural land made up 4.8% of the total land farmed in Germany.

- **Demeter**, the organisation founded by Rudolf Steiner in the 1920s, is the oldest organic grower association and is also known as the BioDynamic Movement. Under this system 1341 farms and 61 817 hectares of agricultural land is currently farmed organically (for more information see www.demeter.de).
- The grower association **Bioland** comprises 4540 farms and a total of 201 959 hectares (for more information see www.bioland.de).
- **Naturland** currently comprises 1784 farms and a total of 82 541 hectares (for more information see www.naturland.de).
- **Biokreis**: 539 farms and a total of 15 521 hectares (www.biokreis.de).
- **Gäa**: 485 farms and a total of 52 872 hectares (www.gaea.de).
- **Biopark**: 667 farms and a total of 134 342 hectares (www.biopark.de).
- **Ecoland**: 17 farms and a total of 623 hectares (www.ecoland.de).
- **Ecovin**: 194 organic wineries and 1098 hectares (www.ecovin.de).

These organic grower associations account for 9606 farming members, and there are another 7185 farms with EU-organic certifications.

Bund Oekologische Lebensmittelwirtschaft e.V. (BOELW) is the new umbrella organisation for agricultural producers, processors and traders of organic food products in Germany (www.boelw.de).

Wholesalers

The original and oldest wholesaler, Rapunzel, has expanded over the last 30 years and now employs 250 staff, with an annual turnover of €70m (see www.rapunzel.de).

Currently, the largest wholesaler is Dennree, which employs 550 staff, has a range of 7500 products and an annual turnover of almost €200m (www.dennree.de).

A further 13 regional wholesalers work in partnership throughout Germany with a total of around 800 employees. With the help of an innovative advertising campaign (produced by www.citrusblau.de) they play a key role in wholesale distribution www.die-regionalen.de. Other smaller wholesalers include Weiling, Ökoring and Epos.

How is organic food distributed?

Conventional retailers lead the distribution of organic products with 37% of annual organic product turnover, followed by more than 2000 organic food stores, which share 26% of the market. Direct marketing and weekly markets account for 16%, health-food shops 8% and bakeries and butchers 7%. The remaining 7% is sold through chemists, mail order and delivery services.

Driving forces

Food scandals such as reclaimed, contaminated or re-used meat, and pesticide residues in vegetables and fruits have been effective in rousing short-lived public reactions but have also helped raise awareness concerning issues surrounding food production and processing. For far too long, big-name companies have stood by and watched as organic food producers outgrew their indigenous markets and developed into large brands themselves within the organic food industry. Conventional food industry retailers and even low-cost supermarkets have recognised the growth potential of organic products and many have successfully introduced their own organic in-house brands to the market.

The successful in-house brand 'BioBio' of Plus (a German discount store) recently showed that organic products can benefit the stagnating food industry. Organic food is a growing market, from which both organic producers and new players in the organic market can profit. Above all, the supply of organic quality raw materials has to keep pace with this increased growth. Big-name companies need reliable suppliers who can provide sufficient quantities, at a good quality. Big businesses do not want to put their reputations on the line, and so thoroughly monitor the origin and quality of their raw materials. The German discount store Aldi has demonstrated that arrangements with producer groups help ensure that the necessary supplier demand is met, and because of this can secure more competitive prices.

Is there room for the big players?

The organic food industry can also profit greatly from big-name companies. They can, through their market size, increase awareness of organic produce and its quality to a large section of consumers who would not otherwise be directly exposed to organic products. Already 50% of the German population occasionally buy organic products.

It is important for the organic food industry to keep pace with the latest developments. Over the short- or long-term, only organic products that can hold their own in terms of appearance and taste will be able to win over consumers.

Although food industry scares and trends such as the Slow Food Movement and natural food are always drawing new groups of consumers into organic food shops, to date organic products account for only 3% of the total food market. The reason for this, according to the organisation 'Foodwatch' (see: www.foodwatch.de) seems to be relatively high retail prices and the fact that organic food is regarded as expensive. Unfortunately, this attitude is well ingrained in consumer consciousness, even though organic products are becoming cheaper. New marketing concepts are able to reduce prices and make organic food more affordable, even for some of the most resistant of consumers.

By the time the first organic supermarkets began to appear, prices of organic foods were already falling and organic products were reaching a wider section of customers. Conventional food industry retailers have been contributing to the increasing acceptance of organic food since 1988, when the trade chain Rewe first placed products from its own in-house organic food brand 'Füllhorn' on its shelves. Nowadays, cashiers at practically all big supermarket chains are regularly scanning organic products at the check-outs.

In 2004, conventional food industry retailers turned over €1.28bn selling organic products, compared with €900m in the specialist organic food industry according to Matthias Horx in Biofach Trendletter, number 3 (www.biofach.de). Even low-cost supermarkets have long been stocking organic food on their shelves. Since early 2005, Aldi-Süd has sold organic fruit and vegetables; in the summer of 2005 Plus expanded its BioBio range to around 50 products and undertook a large-scale media campaign. At the same time Lidl held its first nationwide promotion week for organic products and since the beginning of 2006, the discounter consistently stocks organic products.

However controversial it may sound, this development has contributed more to broadening the organic consumer base than any expensive advertising campaign: it is now commonplace to hear consumers commenting that organic products have become more affordable. Even if this is only true for a few products, available in a particular low-cost supermarket, it nevertheless provides the organic movement with a concrete way forward. The organic movement as a whole should view this as a positive development and should recognise that it will increase the number of consumers buying organic.

Organic products become franchise businesses: Dennree takes new marketing paths

By 1975, Dennree, Germany's largest organic food wholesaler had already brought its first in-house organic brand onto the market. In 2005, Dennree achieved a total turnover of €234m, of which 10% was from its own seven in-house brands. Through partnerships and the first franchise for organic supermarkets in Germany, Dennree has thought its organic food marketing strategy through to its logical conclusion. This particular organic food wholesaler is an example of how lower prices can be achieved without the inevitable compromise on quality. Dennree passes its own savings on to its consumers – thanks to successful marketing campaigns, better logistics and increasing profit margins.

Owing to its Gutfleisch brand, Edeka Nord is also showing that organic meat prices can be substantially reduced when using the company's logistics for both conventional meat and organic products.

Who is the organic consumer in Germany?

The social backgrounds of those who buy organic food are well documented. Of the three main categories – post-materialists, the middle class and the modern performers – over half of the former and a third of the latter two groups regularly purchase organic products. What is not so clear, however, is why different people from the same background choose to buy, or not to buy, organic foods. In a study carried out by the agency Sinus in 2006 (www.sinus-sociovisions.de), a group of women shoppers from each social group was asked their opinions on why they chose or did not choose to buy organic. The results were interesting.

Organic as a matter of taste and principle

The post-material group represent buyers responsible for the initial increase in sales of organic products, and also continue to represent a large proportion of shoppers in independent organic outlets. For this group, organic products are purchased as a matter of principle. Not only out of concern for the environment but nowadays, also equally important are issues of enjoyment, taste, health and well-being. The basic decision to buy organic provides these consumers with the reassurance that they are buying the 'right' products. There are also a few post-materialists, however, who do not buy organic products. Most choose not to for personal negative experiences, and have developed a certain amount of mistrust with regard to organic produce. They tend to deal with their personal concerns about the food industry by buying from and supporting their traditional, regional farmers.

Consumers' motivation

The middle class represent today's class-conscious mainstream who buy organic partly as it appeals to their reason and partly because it's fashionable to do so. They are concerned about pesticide residues in food and want unadulterated products. They are, however, less informed about the organic movement in general, and place their faith in companies and brand names.

This consumer group mostly shops in conventional outlets, but does occasionally strive to shop in wholefood shops. It is important not use extreme marketing tactics in order to reach this group, for fear of alienating them. Members of this group who do not buy organic products, associate organic with healthy living, but at the same time are sceptical as to what products branded as 'organic' actually constitute, and view organic consumers as buying into a craze. Their answer to healthy eating is primarily fresh produce, sourced locally from small independent retailers.

Organic: energy and fitness

The greatest challenge proves to be winning over customers from the third consumer group, the modern performers. This group represents a divergent elite of young people. For them, organic means fitness and energy, this is why they buy into whole-food products. They have little background knowledge of the organic movement and shop primarily in conventional supermarkets because of their close proximity and the wide range of available produce. Therefore wholefood shops find it difficult to target this group, especially as the alternative wholefood image is not in keeping with their lifestyle. The only chance independent retailers have of attracting this consumer group is by offering ample parking facilities and long opening hours.

Marketing insights

The amount of information available to consumers concerning organic products is, on the whole, much lower than experts believe. This may explain the common assumption that organic immediately equals expensive.

Each of these consumer groups has their own tendencies, motivation and 'logic' with regard to organic products and the organic market. Members of the post-material group take responsibility for the environment by choosing to buy organic, yet it is scarcely a consideration for members of the other two social groups.

In relation to the tendencies discussed above, this means that the style of marketing tactics must also target the specific concerns of social groups; otherwise much of the advertising will be to no effect.

It is known that over two-thirds of decisions made whilst shopping are made at the point of sale. As the majority of consumers cannot tell for themselves the difference in quality between conventional and organic products, presentation in the market, i.e. placement, accentuation and communication, is the deciding factor in whether a product sells well.

Media

There are free organic consumer magazines that are distributed throughout 2000 organic food outlets all over Germany. *Schrot und Korn* has a circulation of 634 000 copies and appears monthly (www.schrotundkorn.de). *Eve* has a circulation of 435 000 appearing every two months (www.eve-magazin.de).

There are also two trade magazines: the monthly *BioHandel* with circulation of 7000 copies (www.biohandel-online.de) and the quarterly *Biopress* with a circulation of 15 000–20 000 (www.biopress.de).

Every grower association has its own publication and there are also many environmental and lifestyle magazines that often cover organic issues.

Shows

Germany is fortunate to have the two of the most important trade fairs in the food sector on its doorstep. The leading global organic trade fair BioFach (coincidentally founded by the author Hubert Rottner in 1990) takes place every February with over 2000 exhibitors and 33 000 visitors. This trade fair provides the best overview of what is going on in the organic business world (www.biofach.de).

In addition, every two years the world's largest food show, Anuga takes place in Cologne. In 2005, the event attracted around 6000 exhibitors, 161 000 visitors and around 800 organic stands (see www.anuga.de).

More organic ingredients in restaurants and catering

Most high-profile chefs in Germany use organic produce in their kitchens, and many hospitals, office canteens and student cafeterias also offer a wide range of organic food (see www.biospitzenkoeche.de).

The Slow Food organisation in Germany, which has roughly 5000 members, is still small compared with Italy's 67 000 members, but it is nevertheless successfully promoting seasonable, locally, fresh and organic food through events, visits and dinners throughout Germany. The recommended trade show is Salone del Gusto (see www.slowfood.it).

Natural organic cosmetics

In recent years natural cosmetics have become a major trend. Stars including Julia Roberts, Madonna and other Hollywood beauties use German cosmetics such as Weleda and Wala. There is a growing awareness that we are responsible for the world in which we live, and an increasing concern for our own health. It is important to appreciate the delicate balance of the natural world and to strive to maintain this harmony by using environmentally friendly products that benefit us and cause minimum damage to our environment. Concern for physical appearance must be balanced with finding conscientious means in which to care for our skin and our health in general. Natural cosmetics should offer wholesome personal care in the truest sense (www.weleda.de and www.wala.de).

How does one decide what is good for us and what isn't? Wellness, fitness and an evolving awareness of holistic health have revolutionised the cosmetic market. The choice of products responding to the trend towards natural products is immense.

The website www.kontrollierte-naturkosmetik.de offers an overview of the world of natural cosmetics and skin care products with valuable information that will provide a basis for confident choices.

Fair trade

For 30 years, gepa Fair Handelshaus has stood for trade with a compatible social and environmental impact. The year 2005 with its motto 'Fair forever' was a special jubilee year: gepa celebrated 30 years of fair trade with their overseas partners. Throughout the year, the jubilee was celebrated with special product offers, actions and festivities. Today 'gepa' is Europe's biggest fair trade company, with an annual turnover of more than €33m. Every year, gepa have transferred over €18m for food products, handicrafts and textiles to their business partners in over 170 co-operatives and marketing organisations in Africa, Asia and Latin America. Their products are on sale in 800 world shops and 6000 action groups, but also in numerous supermarkets, organic food shops, business canteens and educational institutions. There is an online shop at www.gepa.de.

Future trends

Biofach.de provides a good service with a half-year future trend letter by Mathias Horx, a trend-explorer from Hamburg (www.horx.com and www.zukunftsinstitut.de).

Outlook

Organic supermarkets are the main drivers of the growing market with 60 new stores in 2005 and a current total of around 300. These stores are opening at a rate of one a week, with a growth of 15–20% per annum. Conventional supermarkets that stock only a basic range of products are under pressure to extend their product range to include more variety to meet the challenge from the discounters who have joined the market with their own cheaper basic products.

Changing consumer demand, more outlets and a growing range of supplies will no doubt lead to rapid sales, further ahead. Remaining challenges include GMOs, the revision of the EU-organic law, unclear governmental support: the most optimistic recent reports are for €20m per year presenting a challenge and shortfall from home-grown supplies. Simonetta Carbonaro, Professor of Humanistic Marketing and a management consultant at Realise Strategic Consultants is quoted as having said in an interview with the financial magazine *brand eins* (Issue 04, April 2006), 'I am convinced that in the age of the Internet, word of mouth is faster and more powerful than any expensive company marketing strategy, though it goes without saying that there has to be a strong message about outstanding value'. This quote brings the chapter full circle, back to where the discussion of organic and fair trade marketing in Germany began, and will no doubt continue.

Reference

1   Willer, H. and Yussefi, M. (Eds) (2006) *The World of Organic Agriculture 2006 Statistics and Emerging Trends Annual Report*. www.fibl.org/shop/index.php

Chapter 11

Organic and Fair Trade Marketing in Italy

Paola Cremonini
Cremonini Consulting

Review of the market for organic food and drink

The Italian organic market emerged in the 1970s during the boom of post-materialistic ideologies: at that time it consisted of a few highly motivated pioneers who had chosen organic food as an aspect of their radical, alternative lifestyles. Their relationship with food was more of a religious matter, which stemmed from two major schools: the macrobiotic (with food being imported mainly from the far-east) and the antroposophic (food mainly imported from Germany). In the 1980s this movement spread to a wider number of consumers, shops, wholesalers and domestic producers, but it was still a niche market: it is only since the 1990s that the market has grown dramatically and experienced a radical transformation, from niche to mass market.

In 1992, when EU regulation 2092/91 assigned organic products their own identity this enabled consumers to recognise organic foods through labelling and certification, even when they were not sold through specialised shops. At the beginning of the 1990s the first organic products were sold through multiples. In the second half of the decade, mainly as a consequence of the BSE scare, consumer awareness of organic food increased dramatically. Within a few years it had spread to the mass market increasing sales exponentially, attracting at the one extreme the mass consumer market – previously totally unaware of organics, and on the other hand, mainstream industries that saw in this fast growing market an opportunity for high return on their investments. In 1999, the first Italian supermarket launched its organic private label range, followed swiftly by other nine major Italian retailers. After the boom years of the late 1990s and the beginning of 2000, sales seem to have reached stagnation. This is not surprising as most consumers switching to organics during the boom years were probably driven by fear of food scares rather than an increased awareness for organic principles. As a result, the mainstream players in Italy have now stopped investing in organic, many have withdrawn from the sector and several supermarkets have decreased the number of organic items on offer (for example, the leading organic supermarket brand 'Esselunga bio' decreased the number of items from 500 in 2004 to 323 in 2005). Nevertheless, statistics indicate that the organic market is undoubtedly an important consolidated market in Italy: in 2005 consumer expenditure for organic food was estimated at €1.5bn (1.5–2% of total food expenditure), the third largest market in Europe after Germany and Great Britain. Italy now represents the third most important extension of organic land in the world after Australia and Argentina, and the first in Europe with more than one million hectares and 40 000 players involved in farming and processing. Italy produces more organic food than it consumes.

In 2006 (starting from end 2005) the specialised sector grew while supermarkets continued to stagnate. In the first eight months of 2006, the organic wholesaler Ecor experienced a growth of 17%. This is partly due to the fact that supermarkets disinvested from the sector and their market share has moved to the specialised trade, and partly to increased efficiency and professionalism of the shops.

In 2005, the distribution and consumption of organic food purchases was mainly through supermarkets; this was because of their generally lower prices and more convenient locations. This sector was followed in importance by the specialised trade represented by 1014 shops. Other outlets were represented by 540 farmers and 659 'agriturismi' (farmers with bed and breakfast facilities) selling directly to consumers, 185 markets taking place on a weekly or monthly base or during special events, and 222 consumer purchasing groups buying directly from producers. A new method being tested in some cities is the fruit and vegetable basket, where groups of farmers deliver fruit and vegetables directly to consumers on a weekly basis.

A growing outlet for organic food in Italy is the catering trade: 275 restaurants in 2005 recording a 10% increase from the previous year. 171 were restaurants, pizza restaurants, self-service restaurants, takeaways, bars and bistros, and 103 were restaurants as part of the 'agriturismi' farms. A highly significant amount of organic food is consumed in school refectories, with one million organic meals per day being prepared in 2005, a 9% increase from 2004. This sector is particularly important for the future growth of the market as it sensitises children to organic food and health issues.[1]

Consumer profile and purchasing motives

In Italy, with a population of over 60 million, 20 million have made at least one purchase of organic food. The typical consumer is between 35 and 44 years old, has a high income and high education level, and lives in northern Italy (including Rome). Seventy-seven per cent of consumers are willing and happy to pay prices that are 5–10% higher, while 54% claim that organic food is restricted in its availability. A survey about the reasons for purchasing organic food, carried out by Biobank in 2005, identified three main categories of consumer profile:

- 47% of consumers buy organic because it is healthy and fosters well-being, is additive and GM-free and is not 'excessively sophisticated'; some consumers have been recommended organic foods by doctors as a remedy for intolerances or allergies and perceive that organic food has a therapeutic effect
- 28% buy organic because it is ethical, respects the environment and is pesticide-free; some consumers identify in their purchase the will to promote and foster organics as an expression of an alternative lifestyle; sometimes this is also associated with a vegetarian diet, or belonging to a consumer purchasing group
- 25% of consumers buy organic because of its intrinsic characteristics: because it is considered better in taste, more nutritious, fresher, genuine, seasonal and because it is diverse: its range offers particular food – such as tofu or seitan – not available in the conventional mass-market

Focus on organic brands

If a brand is sold through the specialised sector it is not usually offered in supermarkets. Two exceptions to this rule are represented by Fattoria Scaldasole in the chilled sector and Alce Nero in the ambient sector, two pioneers that have established their presence from the very beginning of the market development and have strengthened their positioning over the years.

Fattoria Scaldasole

Fattoria Scaldasole consists of a range of yogurt, milk and fruit-based chilled products. As with Ecor – the leader in distribution to specialised shops – Scaldasole's success lies in the strong motivation and belief of its founders in the biodynamic movement (which was launched by Rudolf Steiner in 1924): the founder Mr Roveda from the outset invested in technology and diversification earning the trust of distributors and consumers. Fattoria Scaldasole started in 1986 as a biodynamic dairy farm north of Milan, producing on an artisan level within a closed cycle, distributing products through specialised shops. In the early 1990s it moved to a highly technologically advanced factory and it started to sell to multiples. Throughout the 1990s it diversified its range, including non-organic ambient products such as biscuits. In 1995 it was sold to Plasmon and in 2005 to the French group Andros. Now it is by far the most important organic brand in Italy, present both in the specialised trade and in multiples. Its range includes organic yogurt with and without fruit, some biodynamic certified items, a line of organic yogurt specifically designed for children, organic yogurt with probiotic enzymes 'lactobacillus casei biovitalis' specifically developed by Fattoria Scaldasole, organic UHT milk, organic fresh citrus juices, organic cream cheese and organic cream desserts. Particular care has been taken in developing packaging materials: where plastic pots are used, only the internal part of the pot is plastic, using 50% less material, and the external part is made from paper which can be separated from the plastic for recycling.

Alce Nero

Alce Nero was founded as a co-operative in 1977 in Isola del Piano. Alce Nero – Black Elk – was an American Indian Chief who had fought against Western cowboys. The founders saw a parallel between farming and the American Indian culture, both in harmony with nature, as opposed to the western citizen culture, alienated from nature. Initially a farm was created by a community of young producers around a disused monastery, extending later to a pasta factory. In the 1980s it was one of the first pastas on the shelves of specialised shops. From the 1990s the Alce Nero brand was distributed to all supermarket chains and specialised shops thanks to a sales agreement with a mainstream company. During this time the Alce Nero range grew to include other products such as tomato sauces, extra-virgin olive oil, jams and juices.

At the end of 1999, Alce Nero co-operative and Conapi (Italian Beekeepers and Organic Producers Consortium) created a new company that carried out joint sales, marketing and logistics for both companies. Alce Nero turnover increased from €2.8m to €6m in 2005. Nowadays most of the Alce Nero range is produced by the members of the consortium, which means that total supply chain management is guaranteed. The range includes pasta, honey, jams, biscuits, tomato and fruit preserves, extra-virgin olive oil, rice and other cereals.

Ecor

Ecor is the leading Italian wholesaler of organic food, exclusively supplying specialised shops. It is based in San Vendemiano (near Treviso). It was founded in 1998 from the merger of four pioneering wholesalers: Gea – the biggest, a wholesaler founded in 1987 strongly linked to the anthroposophic movement, La Farnia, Pronatura and Brio. In 2005 it had a turnover of €58m (with a forecasted 2006 turnover of €70m) and 130 employees. It distributes more than 3100 items, ranging from ambient to chilled, frozen to fresh fruit and vegetables, to approximately 800 shops throughout Italy, delivering with its own trucks within 24–36 hours, three times a week.

Ecor's mission is very much linked to Rudolf Steiner's thoughts and philosophy, in that it aims to promote organic food as widely as possible as a means of allowing people to feed themselves in a vital and natural way, while on the other hand it also wants to spread the anthroposophic culture. Of its shares, 70% are owned by the Anthroposophic Society, a non-profit organisation that is actively developing three projects within the anthroposophic movement: the organic (biodynamic) farming co-operative 'San Michele', the kindergarten and primary school 'La Cruna' and the school for the handicapped 'Aurora'.

Ecor's private label, 'Ecor', was established in 2000 and now offers more than 300 commodities. The range comprises ambient food (250 items), fresh (25) and non-food (45). Thanks to Ecor's long-term experience in the organic sector, new products under Ecor's label are constantly being developed, most recently gluten-free, yeast-free and lactose-free product lines. Ecor as a brand has won various national and international graphics and design competitions including the 'Design & Art Direction Annual' (UK), the 'Communication Arts Design Annual' (USA), the 'Grand Prix Design Pubblicità Italia' and the 'Annual Italian Art Directors Club'. Ecor also helps shops by providing know-how on merchandising matters: the B'io project groups together more than 200 shops that follow closely Ecor's directives and marketing strategies, from shop layout to promotional activities.

Naturasi

Naturasi is the biggest franchised chain of specialised shops in Italy with a turnover of €50m in 2004. The first shop was created in 1993 and in 2005 the chain had 43 shops, each with a minimum surface of 300 m², three butcher's shops, two restaurants and one beauty centre; and four shops in Spain. Each shop offers more than 4000 items in ambient, chilled and frozen; 73 items carry Naturasi's label – mainly commodities such as pasta, flour, oil, tomato derivatives, biscuits, jams and juices. Apart from food, Naturasi offers ecologic cosmetics, stationery and housekeeping products.

The founders' idea has been to apply the know-how and marketing tools used in mainstream supermarkets to selling organic and ecological products in specialised shops. Marketing activities are frequent and varied throughout the year: brochures are sent to households, a loyalty card allows the company to gather useful information regarding consumers, and various merchandising techniques are used to increase the visibility of promoted items. Naturasi's own magazine 'InformarSi' is distributed widely within stores, free of charge, and gives information on the products available in store, on organic agriculture, on food and diet in general, as well as information on political and economic issues related to organics and sustainability.

An important project for this company has been fundraising to help the people of Shewula in Swaziland: funds have been raised from 2003 to 2005 through various methods such as gift stamps collections, sales of dedicated calendars, and contributions directly donated by consumers or Naturasi corporation. This has been used for the purchase of GM-free seeds and tools for organic agriculture in the area of Shewula, and for helping a project that hosts 800 children in various communities in Swaziland.

In July 2006 Ecor and Naturasi merged (they bought 49% of each other's shares) with the following goals: to control and co-ordinate the new shop openings (often B'io shops, where opening near Naturasi shops); to create economies of scale in the matter of quality control; to create economies of scale for promoting organic specialised shops and sensitivity to organic food in general.

Multiples carrying an organic private label

In Italy about 1700 multiple outlets offer an organic range. The biggest chains have adopted a private label: Esselunga was the first to launch its own organic private label in 1999 – 'Esselunga Bio' – and now ten supermarket chains altogether have their own brands. According to Biobank [2] in 2005 Esselunga Bio was the leading organic supermarket private label brand with 323 items, followed by COOP with 304, Carrefour with 221 and Rewe Italia with 160. Auchan had 100, Despar 80. All other brands followed with less than 50 items.

Auchan

The Auchan group controls 41 Auchan hypermarkets, 214 SMA supermarkets and Cityper, and 1343 affiliated franchising outlets. It is spread throughout Italy apart from the regions Basilicata and Molise. The organic private label is 'Auchan Bio', which was launched in 2002, and now has more than 160 items. Organic items are either displayed on separate islands or next to conventional and include fruit and vegetable juices and preserves, oils, pasta, rice, flour, drinks, honey, bakery products, milk and dairy products, eggs and chocolate.

Carrefour

The Carrefour group has 43 hypermarkets, 423 supermarkets, 15 hyperstores, 961 DiperDi shops and 18 cash and carry outlets. It sells throughout Italy excluding the Trentino Alto Adige region. Its brands are Carrefour, GS and DiperDi. Its organic private label, 'ScelgoBio', was launched in May 2000. This has over 220 items sold alongside conventional products, ranging from fruit and vegetables to milk and dairy, meat, eggs, bakery, fresh and frozen ready-made meals, pasta, cereals and flours, juices, and fruit and vegetables derivatives, honey and sweeteners, oil and condiments. Carrefour actively promotes its organic line and organic products in general through flyers and catalogues.

Conad

Conad has the brands E.Leclerc-Conad (hypermarkets), Conad (supermarkets and superstores) and Margherita (shops and superettes). It has 17 hypermarkets, 1300 supermarkets and 1300 smaller 'superettes' spread across Italy. The organic label 'Conad Prodotti Da Agricoltura Biologica' launched in November 2000, has 35 items. Its range includes milk and dairy, rice and flour, juices, and fruit and vegetables preserves, bakery products, honey and sweeteners, oils and condiments, and drinks. Organic products are sold either alongside conventional or in dedicated areas. Promotion of organic food is through their in-store magazine or on the internet site.

Coop

Coop has 1276 points of sales, with the brands Coop and Ipercoop (supermarkets and hypermarkets). It is present throughout Italy, excluding Valle D'Aosta, Sardinia and Calabria. It was one of the first to launch an organic private label, in November 2000 under the name 'bio-logici Coop'. The range has 304 items and includes fruit and vegetables, milk and dairy, meat, eggs, baby food, pasta, rice, cereals and flour, juices and preserves from fruit and vegetables, drinks, bakery products, honey, sugar, oils, frozen foods and garden seeds. Fruit and vegetables are sold in a separate dedicated area, whereas other products are sold next to conventional. Prices of fruit and vegetables are 25–30% higher than conventional, milk and derivatives 0–10%, baby food 10–15% and fruit and vegetable preserves 10% higher. Coop is very active in the promotion of organic food with articles in its magazine, brochures, campaigns and sponsorship.

Crai

Crai is a group of independent supermarkets present in all regions apart from Tuscany, Umbria and Molise. Its organic private label 'Crai Bio' was launched in November 2001 and has 39 items: eggs, bread and bread substitutes, pasta, cereals and flours, milk and products, juices and preserves, honey and sweeteners, oil and condiments. Organic products are shown next to conventional with prices 30% higher than conventional. Promotional information on organics is distributed in the stores.

Despar

Despar is a group of approximately 200 superettes, supermarkets and hypermarkets in Trentino-Alto Adige, Veneto, Friuli Venezia Giulia, Emilia Romagna, Lombardia, Tuscany, Umbria and Campania. The organic private label brand 'Bio Logico' was launched in February 2001 for fruit, vegetables, milk and dairy, eggs, fresh juices, juices and fruit and vegetables preserves, pulses, bakery products, oils, pasta, flour and cereals. Organic fresh fruit and vegetables are sold in a separate dedicated area, and the other products next to conventional. Organic products prices are 30% above conventional for fresh fruit and vegetables, and 10–15% for other products. Information on organics is given through promotional campaigns, brochures in the stores and signalling of products on shelves.

Esselunga

Esselunga owns 129 supermarkets and superstores in Lumbardy, Veneto, Piedmont, Emilia Romagna and Tuscany. 'Esselunga Bio' is a leading supermarket brand in organic food. It was the first organic range in November 1999, and now it has over 320 items, including fresh fruit and vegetables, milk and derivatives, eggs, bread, chilled and frozen ready-made meals, baby food, pasta, cereals and flours, juices, and fruit and vegetable preserves, bakery products, honey and sweeteners, drinks and pulses. Organic products are shown next to conventional except for fresh fruit and vegetables, which are sold in a separate area. Price differences in comparison to conventional are 30% higher. Organic products are identified on shelves. The organic private label 'Esselunga Bio' was promoted extensively in its early years, having been launched with an extensive poster-based promotional campaign, which generated high consumer awareness of this particular brand and helped the development of the market in general.

Gruppo PAM

The PAM group is a group of 18 hypermarkets, 105 supermarkets and 66 superettes located in northern and central Italy. Brands are Panorama, Pam and Metà. 'Biopiu' is the organic own-label that was launched in 2000 for fresh fruit and vegetables. and only later expanded into ambient lines.

Rewe italia

Rewe Italia has brands Billa, Standa Supermercati and IperStanda with 168 supermarkets and 20 hypermarkets in all regions apart from Sicily and Molise. The organic private label is 'Si!' Naturalmente launched in December 2001, which includes more than 160 items: fruit and vegetables, pulses, dry fruit, milk and dairy, meat, eggs, bread and substitutes, oils and condiments, wine, drinks and sugar. Fruit and vegetables are displayed in a dedicated corner, whereas other products are shown next to conventional. Prices are 30–40% higher than conventional for fruit and vegetables, the same for milk and dairy, 20% higher for ready-made meals, and 10% higher for wine.

Selex

The Selex group has more than 1000 hypermarkets, supermarkets and superettes with brands A&O, Famila, Dok, Pan, Galassia and Emisfero. The organic private label 'Bio Selex' was launched in May 2001, and includes 19 items: eggs, juices and preserves from fruit and vegetables, oils and condiments. Products are displayed next to conventional products.

Review of the market for fair trade food and drink

In 1994, the Italian certification body Transfair (a member of the Fairtrade Labelling Organizations International) was established. Since then trade has increased dramatically and in 2004 more than eight million Italians had bought at least one fair trade item. Nowadays the fair trade market involves a complete network of importers, processors, distributors and shops with more than 550 companies worldwide certified by Transfair. Italy represents one-sixth of the world's total turnover in the fair trade sector and over one-third of the turnover in Europe. The main focus of the trade is food, followed by handicraft and textiles. Most processed food products are packed in the country of origin, with a smaller proportion being imported as raw materials and further processed and packed in Italy. Main items are bananas, pineapple, coffee, sugar and cocoa.

Sales to consumers are through supermarkets, organic and conventional shops, chemists (with over 4600 points of sales), and a network of approximately 500 shops specialising in fair trade sales, called Botteghe del Mondo. The three main brands are Altromercato, Equosolidale and Libero Mondo.

Coop Italia

Coop Italia is the leading supermarket chain for fair trade food distribution: it distributes its own range labelled 'Solidal' certified by Transfair. Coop acts as a real promoter of the fair trade market, because on one side it has direct contacts with farmers and suppliers abroad, and on the other it carries out extensive promotional campaigns to inform consumers about projects and introduce them to the ethics of fair trade. Coop's magazine 'Consumatori' gives information on the Solidal range and on particular projects. The Solidal range was awarded the Ethic Award 2005, from the leading retail magazine *GDO Week* to reward the best fair and sustainable projects from the retail sector.

The Solidal range includes the following items:

- coffee from Guatemala, Colombia, Peru, Bolivia, Mexico and Chiapas
- tea from India and Sri Lanka
- chocolate bars, chocolate eggs and chocolate cream from the Dominican Republic, Costa Rica, Ecuador, Belize, Ghana and Paraguay
- honey from Mexico/Chiapas, Guatemala and Nicaragua
- cane sugar from Peru
- bananas from Ecuador
- pineapple from Costa Rica and Ghana
- orange juice from Brazil
- rice from Thailand

Botteghe del Mondo

The network of fair trade shops Botteghe del Mondo ('shops from the world') has about 500 points of sale across the country. It distributes not only foods but also handicraft and textiles. About 150 groups of small producers in Asia, Africa and Latin America are involved in the production of items sold directly to the Botteghe del Mondo network through the platform CTM Altromercato. Often shops organise activities on an individual basis aimed at informing consumers, and they also act as meeting points for people sharing common values on sustainability issues.

CTM Altromercato

CTM Altromercato started in 1988 in Bolzano as a co-operative under the name CTM Cooperativa Terzo Mondo (Third World Co-operative). In 1998, the name was changed to CTM Altromercato ('other market') as it became the importing platform for Botteghe del Mondo. Its brand 'Altromercato' is also sold successfully in supermarkets. The co-operative employs 25 workers and more than 3000 volunteers in the Bottheghe del Mondo shops. It coordinates 150 groups of small workers in Asia, Africa and Latin America. Products include coffee, tea, honey, herbal teas, sugar, cocoa, jams, biscuits, cereals, snacks, nuts, chips, cookies, chocolates, candies, juices, syrups and drinks.

Commercio Alternativo

Commercio Alternativo is an importer and distributor of fair trade items with the brand Equosolidale. It is a non-profit organisation based in Ferrara, founded in 1992. It currently imports goods from more than 50 producer groups based in Africa, Asia and Latin America. Products include coffee, tea, herbal tea, barley, cocoa, rice, spreads, sauces, chips, sugar, chips, candies, chocolate, snacks, biscuits, honey, jams, desserts, oil, spices, juices, wine and nuts.

Libero Mondo

'Libero Mondo' is a co-operative established in 1997 from an association of volunteers, 'Tsedaqua'. It has 103 members and acts both as an importer and processor. It imports from approximately 30 countries and 80 producers groups, both food and non-food products and distributes to Botteghe del Mondo and organic food shops. On the processing level it produces biscuits and pasta on a small scale using fair trade ingredients. Moreover, Libero Mondo is also involved in public relations activities towards universities, shops and consumers to promote and develop fair trade concepts.

References

1   Mingozzi, A. and Bertino R.M. (Eds) (2006) *Tutto Bio 2006 Annuario Del Biologico*. Egaf edizioni, Forli, Italy.
2   Biobank (www.biobank.it)

Webliography

Press centre for SANA website press releases can be found here:
http://www.sana.it/sanabologna/ing/sanabo/media/prestampa.html

SANA (2005) Alleanze per successi sostenibili (press release). www.sana.it

SANA (2005) Equo-solidale, consumi politically correct (press release). www.sana.it

SANA (2005) Le dimensioni dell'agricoltura biologica (press release). www.sana.it

SANA (2005) Alimentazione – Scheda dati mercato 2004 (press release). www.sana.it

Chapter 12

Organic and Fair Trade Marketing in the USA

Elaine Lipson
Organic Program Director, New Hope Natural Media, and author of *The Organic Foods Sourcebook*

Introduction

Two divergent trends, complementary at best and antagonistic at worst, characterise the organic foods market in the USA as this book goes to press in 2006. The first of these is the mainstreaming of organic foods; long in evidence, this trend is accelerating rapidly as large conglomerates continue to acquire organic brands, mainstream supermarkets embrace organics, Wal-Mart makes a much-publicised commitment to increase organic offerings, and successful natural/organic supermarket chains such as Whole Foods reveal ambitious growth plans.

The second trend is the emergence of a much smaller but intriguing market, sometimes called 'beyond organic,' that positions itself alongside, or perhaps in reaction to, the mainstream organic market. In this sector, the organic label may be secondary to other value-added attributes in the marketplace, such as local or regional production, fair trade certification, humane animal treatment standards, family farm sourcing, grass-fed/pasture-raised livestock, or other production or social responsibility characteristics that are not comprehensively covered in organic foods standards. In some cases, these products eschew the organic label and its federal requirements altogether, preferring to communicate their practices directly to consumers; in others, organic production standards act as a baseline that these additional distinctions enhance.

A running dialogue in the food, farming and business communities links these two trends, as stakeholders debate the 'true' spirit and correct direction of organic foods and agriculture. This debate played out in 2005 in the regulatory arena, when a small farmer named Arthur Harvey took his disappointment with organic standards to court, at one point winning a victory – later overturned – that might have disrupted the existing juggernaut of organic commerce in the USA but reclaimed the purest interpretation of the organic label. Harvey's efforts were cheered by those who felt that organic standards have been compromised in favour of large multinational companies. He was opposed with equal vehemence by many of the largest organic foods companies and the Organic Trade Association, which successfully undertook a legal initiative to overturn the Harvey ruling, despite a rush of bad publicity and consumer opposition.

The tension between advocates of 'big organic' and 'small organic', or 'corporate organic' and 'beyond organic', is still in play, with a focus now on standards for dairy cows and on the implications of Wal-Mart's organic initiative. A growing advocacy movement for locally produced foods is calling into question the wisdom of an organic market whose success depends on imported ingredients and energy-intensive 'food miles'. As the dialogue continues and the market evolves, it's clear that organic foods and farms are here to stay in the US market.

Both the mainstreaming and what might be called the authenticating trends may well continue to develop with great momentum, as mainstream organic brands become a baseline for healthy and environmentally safe eating, while awareness of attributes that fall outside USDA organic standards grows among those consumers who are most conscious of values.

In a best-case scenario, growing consumer interest in and passion for locally produced foods and high standards will encourage larger, mainstream brands to be farmer-friendly and to be more than just minimally compliant with US Department of Agriculture (USDA) regulations, and to act in the true environmental and sustainable spirit of the organic movement.

Projected growth of the organic market

The rapid growth of the organic market makes it clear why conventional food companies are eager to acquire organic brands and gain entry into the sector. The US retail market for organic foods was estimated to be approximately $13.8bn in 2005, according to the Organic Trade Association 2006 Manufacturer Survey conducted by *Nutrition Business Journal*, and is projected to reach $15.9bn in 2006.

This represents 16.2% growth in the market in 2005 over 2004, and 15% growth in 2006 over 2005. Historically, the market has had double-digit growth for many years, as Table 12.1 illustrates, compared to growth in the conventional food industry of 2–3% annually.

Table 12.1   US organic sales, 1997–2006 (in millions, $US).

	1997	1998	1999	2000	2001	2002	2003	2004	2005	2006 *
Total organic	3590	4290	5040	6100	7360	8640	10380	11900	13830	15910
Per cent growth		19.5	17.5	21.0	20.7	17.4	20.1	14.6	16.2	15.0

* 2006 figures projected
Source: Organic Trade Association 2006 Manufacturer Survey, conducted by *Nutrition Business Journal*.

The report projects that organic foods will make up about 2.8% of the entire US food industry in 2006.

High growth organic categories

Fruit and vegetables comprise the largest category of organic foods in the USA by sales dollars, with $5.4bn in 2005. Dairy products follow with $2.1bn (see Table 12.2).

Though meat is the smallest category in sales dollars, it is also the fastest-growing category, increasing 55.4% from 2004 to 2005. Perhaps more than in any other category, organic meats have hurdles to overcome in both price point and consumer understanding, because the natural meat category is well established and quite competitive to the organic meat market. The primary points of differentiation between organic and natural – organic feed and third-party certification – are abstract to many consumers, who may be satisfied with labels that claim that no antibiotics or growth hormones have been administered, and that the animals have been given vegetarian feed. Natural meat brands vary widely, from those with great integrity to those that are just a notch above conventional meats, so companies must strive for greater consumer education and strong brand identities.

That said, the organic meat category may be especially poised to take advantage of food scares – not a particularly sustainable or reliable growth plan, but one that's driven organic growth before. Eric Schlosser's best-selling book about industrial meat and fast food production, *Fast Food Nation*, has been made into a film with a release date of October 2006. If successful, the movie may do more for alternative meat producers than even the threat of mad cow disease in the USA, which fuelled the first big wave of growth in the organic meat category.

Growth in this category has a direct impact on other categories. Organic livestock must be given certified organic feed, so category growth is creating high demand

Table 12.2 Organic foods sales in 2005 by category/percentage of conventional foods sales (in $US millions).

Category	Organic sales	Organic growth (%)	Conventional sales total	Organic %
Dairy	2140	23.5	62006	3.45
Fruit and vegetables	5369	10.9	95905	5.60
Breads and grains	1360	19.2	62.001	2.19
Meat, fish, poultry	256	55.4	117.422	0.22
Beverages	1940	13.2	105262	1.84
Snack foods	667	18.3	33290	2.00
Packaged/prepared foods	1758	19.4	61440	2.86
Condiments	341	24.2	19464	1.75

Source: Organic Trade Association 2006 Manufacturer Survey, conducted by *Nutrition Business Journal*.

for this commodity. Organic livestock also require grazing land that meets organic standards; increases in organic acreage in the USA can largely be attributed to this category. Finally, there is a growing market for pet foods made with organic meat, allowing livestock producers to make full use of livestock and create more profitable and efficient systems.

Challenges to the organic dairy sector

Organic dairy products have seen tremendous success in the USA, growing quickly into one of the largest and fastest-growing categories in the organic marketplace. Recently, however, demand has outpaced supply, creating a scenario where dairy companies compete to contract with dairy farmers and are developing stronger programs to assist conventional dairy farmers in transitioning.

At the same time, the watchdog groups Organic Consumers Association and Cornucopia are publicly criticising what they say is factory organic dairy production, or 'feedlot organic', by large producers with thousands of head of cattle together in a confinement situation. Horizon Organic, owned by Dean Foods, has been the most publicly criticised brand; in response Horizon says that 80% of its milk comes from small and family farms, that it is actively assisting more small farmers in transitioning to organic, and that it is transitioning more pasture land so that its large Idaho dairy farm cows have more access to pasture. Horizon has also said that it supports strengthening of pasturing regulations by the USDA.

The USDA will need to address livestock pasturing as well as the regulations for replacement animals in dairy herds. Changes in language related to the Harvey v. Veneman legislation of 2005 must be clarified with regard to using conventional animals that have been transitioned to organic, versus animals raised organically from birth.

The controversy provided marketing opportunities for smaller brands and those that are farmer-owned, such as the Organic Valley co-operative. Consumers are learning more about how dairy cows are raised and treated, and scientists are examining the benefits of meat and dairy products from cows raised on pasture and those raised on feed. The USDA is expected to release new proposed pasturing regulations in 2006.

The quandary of imported ingredients

Continued rapid growth for organics depends on an adequate supply of ingredients and ongoing certification of new acreage. Manufacturers are increasingly fuelling the rapidly growing demand for organics by importing ingredients into the USA rather than helping to finance and build greater domestic organic acreage. According to

USDA, more than $1bn is being spent annually on importing organic products to the USA, with an import-to-export ratio of eight to one.

According to the Organic Trade Association 2006 Manufacturer Survey, conducted by *Nutrition Business Journal* (*NBJ*), 23% of survey respondents said that they sourced no raw materials offshore, and 21% said they sourced 91–100% of organic raw materials from sources outside the USA. In its March 2006 issue, *NBJ* stated, 'The 194 respondents to this question yielded an aggregate 38% of materials sourced from overseas.' Manufacturers expect that a consistent supply of quality organic ingredients and raw materials will be their greatest challenge in the expanding market.

Organic consumers may be increasingly critical of widespread use of imported organic ingredients, especially when they are sceptical of the integrity of production and certification practices in exporting countries. A growing awareness of the social and economic benefits of local and regional food sourcing is fuelling this dialogue in the USA. Though the environmental implications of food miles are perhaps less well understood by consumers in the USA than in the UK, the concept of local food systems is compelling nonetheless.

But the distance that organic food travels from field to table is not specifically a part of the organic standards. In fact, the national organic standards were intended in part to facilitate global trade. Scale of farms, labour and fair trade issues, and specifics of animal treatment are also not areas that are well defined by USDA. An understanding of the US regulations for the organic label is essential for making sense of the current landscape, as well as prospects for selling and marketing into the US market.

The development of USDA organic standards and labelling

The United States Congress passed the Organic Foods Production Act (OFPA) in 1990, mandating the creation of a uniform federal set of standards for organic foods and giving regulatory oversight to the USDA. The passage of OFPA was unusual in that the organic industry actually sought federal oversight. As organically grown foods grew in popularity, there were occurrences of fraud as it became apparent that consumers would pay more for products they believed to be grown with organic methods. Faced also with hostility from the food industry, it made sense to exchange the patchwork certification structure that had emerged for a consistent definition of organic that would, in theory, protect both farmers and consumers from the consequences of fraudulent use of the label.

OFPA also mandated a citizen advisory board that would work closely with USDA to create organic standards and monitor their evolution. It took OFPA and USDA a full decade to create a rule that the organic community, including organic consumers, accepted. The final organic rule, at 600-plus pages, was released in May 2001, with an implementation date of October 2002.

In broad strokes, USDA standards for the organic label prohibit the use of:

- toxic and synthetic pesticides, herbicides, fungicides, fertilisers and preservatives
- transgenic technologies, or genetically modified ingredients
- irradiation
- processed sewage sludge as fertiliser
- antibiotics in livestock
- growth hormones in livestock

In addition, animals must be given organic feed, and are required to have access to pasture and humane treatment.

All organic products using the organic label must be certified by a USDA-accredited independent third party agency. Most of the existing certification bodies operating before 2001 are now accredited by USDA, including California Certified Organic Farmers, Oregon Tilth, Northeast Organic Farming Association, and many state departments of agriculture. Processing facilities must also be certified. Retailers are not required to be certified unless they are also operating as handlers, but some stores have sought certification, including the Whole Foods Market chain.

In general, land must be documented as free of chemical applications for three full years before organic certification can be granted. Other requirements for organic certification include a farm systems plan and ongoing, comprehensive documentation of farm practices.

Four categories of organic labels

USDA organic standards provide for four categories of organic foods. The labelling scheme, with its different levels of organics (see Table 12.3), was designed to accommodate processed foods where some non-organic ingredients would have to be used. In terms of catalysing industry growth, this structure has been successful, but in terms of consumer understanding and legal clarity, perhaps less so, as we'll see.

National Organic Standards Board

The Organic Foods Production Act of 1990 mandated, along with the creation of federal standards for the organic label, the creation of an ongoing citizen advisory board that would help guide USDA decisions as the organic market evolved. The 15-member National Organic Standards Board (NOSB), includes farmer, handler,

Table 12.3 Categories of organic labels.

Category	May use USDA organic seal	Description	Labelling limitations
100% organic	Yes	Must consist of 100% certified organic ingredients.	None – may identify product as 100% organic
Organic	Yes	Must consist of 95% or more certified organic ingredients. There are restrictions on allowable ingredients and production methods in the remaining non-organic 5%.	None – may identify product as organic
Made with organic Ingredients	No	Must consist of 70% or more certified organic ingredients. There are restrictions on the remaining non-organic 30% ingredients.	May claim up to three "Made With Organic…" ingredients on the main display panel.
Less than 70% organic	No	May list the actual percentage of organic ingredients (i.e. 36% organic ingredients) but not on main display panel.	May only list organic ingredients on the side panel.

Source: USDA National Organic Program.

certifier, manufacturer, scientific, retailer, environmental and public interest representatives.

The creation of the NOSB was hailed as a key element in the unique public/private partnership that the federal organic standards were intended to be; however, during the controversy over the Harvey v. Veneman lawsuit of 2005, some in the organic community were concerned that the changes written into the standards (to override the Harvey decision) could weaken the role of the NOSB. It is an advisory board only, charged with making recommendations to USDA, but is an important link in voicing the interests of a wide swath of stakeholders in the organic community.

Each NOSB meeting includes public comment periods. Information about meetings, as well as transcripts of previous meetings, is available at /www.ams.usda.gov/nosb/index.htm.

Harvey v. Veneman and changes to USDA organic standards

Much of the controversy in the USA about the future of the organic movement and industry centres on aspects of farming and food production that are not covered in the organic standards. These include scale and size of farm operations, specific rules for humane animal treatment, specific rules concerning grazing and access to pasture for livestock, fair trade and living wage standards, and issues concerning replacement animals in dairy herds.

Because of the legal manoeuvres that centred on the aforementioned Harvey v. Veneman lawsuit in 2005, consumers also learned more about the presence or possibility of synthetic ingredients in organic food products. USDA regulations provide for a *National List* of approved synthetic and prohibited natural substances; at present 38 synthetics are approved for use in organic production. A survey conducted at the height of the Harvey controversy by Consumers Union, a non-profit consumer advocacy organization based in Yonkers, New York, that publishes *Consumer Reports* magazine and the online site www.eco-labels.org, found that most consumers believe there are no synthetics allowed in organic foods. This isn't too surprising, since throughout the history of organic food standards, consumer understanding of both their protections and limits has been mixed at best.

Critics of the Harvey ruling predicted that it would result in many more synthetics being allowed in organic foods, especially as 'food contact substances' that are not listed on the ingredients panel of processed foods. Since the federal rulemaking process will take some time, the long-term resolution of these issues remains to be seen. Katherine DiMatteo, the Organic Trade Association's former executive director, told *Natural Foods Merchandiser* in its December 2005 issue that food contact substances had yet to be explored in relation to organics. 'We do need to have discussions about food contact substances, and right now I don't think that any of us even have information about that to know the solution. We need to start [by learning more], and then begin to have the conversation about whether they are appropriate or inappropriate, and whether as a whole group they should be included or excluded,' she said.

Although the non-profit Organic Consumers Association, which led opposition to the Harvey ruling, claims that more than 300 000 consumers called, wrote, or emailed Congress to speak out against the legislation, retailers and manufacturers say that there has been no negative impact on sales since the amendment was passed. Quite the opposite, in fact – the organic movement seems to be growing apace, with more mainstream food industry players eager to get involved.

Anecdotally, however, conversations with individual consumers suggest that there is some erosion of trust in the standards, some disillusionment among those who have been supporting organic for many years. Because these devoted organic shoppers are not likely to return to conventional foods, they may still be buying organic, perhaps with more awareness of brand ownership (many organic brands downplay their ownership by food industry giants). The Harvey lawsuit controversy has been replaced by discussions of pasturing dairy cows, and in particular, of Wal-Mart's embrace of organic products, but the overarching questions remain. What has organic become, in what direction is it headed, and how much has the spirit of the organic movement been compromised by industry success?

Sales outlets for organic foods in the USA

According to *NBJ* research derived from the Organic Trade Association 2006 Manufacturer Survey and other sources, about 48% of organic food sales in the USA take place in natural food and specialised retail outlets, including the Whole Foods, Wild Oats and Trader Joe's chains, as well as many independent natural products retailers. Of organic sales, 45% occur in the mass market, including mainstream supermarkets and discount chains such as Costco and Wal-Mart. About 7% of organic foods sales are direct to consumer through farmers' markets, community supported agriculture (CSA) schemes and farm shops.

Whole Foods Market

Whole Foods is the largest chain of natural and organic foods supermarkets in the world. In 2005, the company had $4.7bn in sales, with 184 stores in the USA, Canada and the UK (a flagship London store is scheduled to open in 2007). The company has stated a goal of $12bn in sales by 2010. An aggressive expansion strategy by Whole Foods has included both new stores and acquisitions, such as the Fresh & Wild chain in London. Whole Foods has more than 30 000 employees, which it calls 'team members' – a cornerstone of the Whole Foods culture.

Though Whole Foods does not sell only organic foods, it was the first national chain to opt for voluntary organic certification under USDA regulations for its adherence to correct retailing practices with organic foods. Though the company has been criticised for offering too many conventional fruits and vegetables in its produce departments, it arguably sells more organic foods than any other entity. Whole Foods has also been criticised in the past for distribution systems that create barriers to entry for local foods, but recently announced a new company-wide commitment to sourcing locally. Whole Foods founder John Mackey remains deeply engaged with these and other food issues, and keeps a blog on the company website (www.wholefoodsmarket.com/blogs/jm) where he responds to some of the criticisms levelled at the chain.

Whole Foods Market has a unique company culture that it says contributes as much to its success as its beautiful and abundant stores. Pay scales for the best-paid employees are scaled to those who make the least; teams and team members are empowered to make decisions in matters that affect them and their stores. The company opposes unionisation and has repeatedly blocked attempts by some employees to unionise; the company repeatedly appears on lists of the best places to work in the USA.

In a July 2004 profile of John Mackey in *Fast Company* magazine, journalist Charles Fishman wrote, ' ... Whole Foods creates markets that are a celebration of food: bright, well-staffed and seductive; a mouth-watering festival of colors, smells, and textures; a homage to the appetite.' Communities around the country lobby for Whole Foods to open stores, confident that it will bring economic advantage and happy customers.

Whole Foods has made investors and Wall Street happy too, thus helping to legitimise organic foods and many of the positions that the company advocates. After the recent Wal-Mart announcement that it would offer more competitively priced organic foods, Whole Foods stores in the New York area advertised a new campaign of lower-priced organics. Combating their 'Whole Paycheck' nickname may necessarily be a part of their strategy going forward, as others compete for their customer base.

Wild Oats

Though once highly competitive with Whole Foods, Wild Oats has not met with the same success as a publicly traded company in the past several years. With 110 stores in 24 states and the Canadian province of British Columbia, Wild Oats claimed more than $1bn in sales in 2005. Wild Oats has suffered from a more fragmented and less cohesive identity than Whole Foods, aiming for a more 'alternative' image but also populating its leadership team with executives from conventional food industry backgrounds without a personal history with the company, the natural foods industry or the organic movement.

Wild Oats has also had difficulties with its primary distributor in recent years, forcing a switch back to a previous partner. Other challenges have dogged the company, and expansion strategies have met with limited success. The company has long been rumoured to be a ripe candidate for acquisition.

With stock prices for Whole Foods out of reach for many, however, Wild Oats remains a target of interest for investors who are optimistic that the company can turn around and that interest in the organic market is so strong that opportunities with this company remain viable.

Wild Oats is scheduled to open a new flagship store near its company headquarters in Boulder, Colorado, in 2006 – a stone's throw from one of the busiest stores of Whole Foods Markets, and which also has plans for expansion.

Trader Joe's

Privately held Trader Joe's had estimated sales of $3.2bn in 2005 with 230 stores in 19 states. With an approach built on selling gourmet foods at great prices, Trader Joe's offers many organic items, but with less of an organic mission-driven philosophy than Whole Foods or Wild Oats. This sits just fine with their customers, who cherish the low prices, down-to-earth feel of the stores and smaller-size stores (about 10000 ft^2 compared to stores of 35000 ft^2 and 80000 ft^2 or more for Wild Oats and

Whole Foods, respectively). Trader Joe's stock in trade is private label; the company has more than 2000 private label products, and many of them are organic.

Independent natural foods stores

The backbone of the natural products movement is the thousands of smaller independent natural foods stores and small chains that exist in just about every community in America. Some of the better-known small chains include Vitamin Cottage in Colorado, New Leaf Community Markets in northern California, Earth Fare in North and South Carolina, Puget Consumer Co-Op (PCC) in Washington state, Mustard Seed Market in Ohio, Open Harvest in Nebraska and Mother's Market in southern California. Most of these stores have deep roots in their communities and in the organic movement.

It can be challenging, however, for smaller stores to compete when natural foods supermarkets come to town. Renewed interest in local foods and local businesses can help here, as well as creative marketing strategies. Just as small farmers must successfully market their brands to stand out, so too must small independent stores.

Mainstream supermarkets

As the organic market matures, mainstream supermarkets are rapidly gaining a market share of organic foods. Chains such as Kroger's and its affiliates, Safeway, Albertson's, Von's, Hy-Vee, H.E. Butts' Central Market, Wegman's, Ralph's and others have all created strong initiatives for organics, including extensive private label lines. Although it seemed for many years that these stores were reluctant to showcase organic for fear of 'making the other food look bad,' or something to that effect, conventional and organic foods now coexist on supermarket shelves without conflict. Of course, it helps that many of these brands are now owned by the same industrial food giants that make conventional foods, such as Kraft Foods, General Mills, Kellogg, Dean Foods and Coca-Cola.

Stores differ in their approach to merchandising organics, with some opting for separate natural foods sections and others integrating organic and natural choices on the shelf next to their conventional counterparts.

Mass-market stores and Wal-Mart

Some organic offerings have been available in large mass-market discount stores such as Target, Sam's Club, and Costco, and yes, Wal-Mart, for years. National brands like Horizon milk and Earthbound Farm packaged salad mixes have done well, and packaged organic foods such as pasta sauce have been easily integrated into these stores. Wal-Mart's announcement in early 2006 that they intended to double their organic items and price them with only about a 10% premium over conventional foods, however, set the organic world on its head, and months later, on a Sunday in June 2006, it continued to be the topic of feature story in the *New York Times Magazine* (while, on the same day, the CBS television show *60 Minutes* profiled Whole Foods founder John Mackey).

Wal-Mart is the biggest retailer in the world, and sells about 14% of all the groceries in America, according to *USA Today*. Critics claim that it puts tremendous price pressure on its suppliers, that it drives smaller stores out of business, and that it shaves its employee benefits to the bone. Certainly Wal-Mart caters to America's most conservative bent, and has censored or refused to sell magazines or music that it deemed inappropriate.

Aside from philosophical and cultural traits that make it seem like an uneasy marriage with organics, Wal-Mart requires enormous inventory of the brands it chooses to sell. When it comes to organics, then, the big question is: Where will it all come from? How fast can manufacturers ramp up production, with a three-year transition period required for farms to be certified organic? And how will they be able to cut prices to meet Wal-Mart's terms? In the end, will the pressure to increase supply and lower prices result in more attempts to weaken standards, and/or in an organic market where cheap organic ingredients are imported from China?

No-one can or should dictate who sells organic foods, and who has access to them. For millions of Americans, Wal-Mart is their primary or only grocery store. Shouldn't these shoppers have as much access to organics as the urban dweller with Whole Foods or Wild Oats nearby, organics in their local Safeway, a farmers' market every Saturday and independent natural foods stores dotting the landscape?

In an ideal outcome, the Wal-Mart initiative will help more farmers to make the transition to organic, and perhaps will help Wal-Mart embrace the principles of the organic movement that still matter to so many people – strong standards, transparency, a fair price for food that doesn't hide externalised costs, and a response to ecosystems that honours their unique characteristics rather than trading down to the agribusiness model. It will be difficult for Wal-Mart or any other entity to put pressure on organic standards without notice; whether or not the organic community is unified enough and organised enough to direct its own future is another story.

Though the Costco chain often gets categorised with Wal-Mart, its corporate culture is very different; where Wal-Mart seems like a situation fraught with uncertainty, there seems to be less discomfort among organic advocates with Costco and other of the so-called big-box retailers. In totality, it's clear that supply of organic ingredients and products will be a challenge for many manufacturers; competitive pricing is likely to be an issue; and we must hope that these retailers are not in this market for the short term, but will work with manufacturers and growers to create sustainable partnerships.

Farmers markets and CSA farms

At the other end of the spectrum, farmers' markets and community supported agriculture (CSA) farms that sell directly to consumers are thriving in the USA. CSA farms sell memberships, or shares, in advance of the growing season. Eaters therefore become partners in the farms' risks and successes. Generally, CSAs operate by distributing boxes of fresh produce to members each week. The contents of the box vary, of course, depending on the stage of the season and weather. Shoppers might find unusual produce that they wouldn't normally purchase. Some CSAs include flowers, eggs, herbs and other farm products, as well as recipes and newsletters.

The number of farmers' markets has grown significantly in the USA in tandem with the growing organic marketplace and awareness of farms and local foods. Although not all farmers selling direct are organic, and many small organic farms selling direct to consumer opt out of certification (farms selling less than $5000 annually are not required to be certified, and may still call their product organic as long as they are in compliance with USDA standards), they offer the consumer the ideal situation of being able to talk directly with the producer about farm practices.

Fair trade in the USA

Organic standards in the USA, as has been noted, do not cover in any detail practices that fall under the heading of social responsibility. It was perhaps assumed that organic principles would encompass issues such as treatment of farm workers, and that the market would dictate prices paid to growers. With time, consumers and advocates have identified more clearly the areas in which the organic standards can be enriched by additional criteria to meet more complete notions of sustainability. The fair trade movement has arisen to address the need for a guarantee of fair and appropriate payments to growers, especially small producers in the third world where exploitation and price controls are common.

Consumer consciousness of fair trade is less developed in the USA than in the United Kingdom and elsewhere, but awareness is increasing with the number of products and commodities claiming fairly traded status. As with organic, manufacturers engaging in fair trade practices often seek third-party certification to a codified set of standards as a guarantee to consumers.

Transfair USA is the primary certification agency for fair trade products sold in the USA. Founded in 1996, Transfair USA is one of 17 members of Fairtrade Labelling Organizations International, a global body based in Bonn, Germany. Transfair's criteria for the 'Fair Trade Certified' label include small-scale production, long-term relationships, a minimum threshold price paid to growers, a fair price based on the marketplace and credit guarantees for growers. The Fair Trade Certified label is rated as 'highly meaningful' by www.eco-labels.org, a website that evaluates eco-labels and claims published by Consumers Union, a respected nonprofit consumer advocacy organisation.

The Fairtrade Certified label appears on, and is best understood by US consumers on, coffee, chocolate, bananas, sugar, and increasingly, tea, other tropical fruits, and rice. Many, but not all, fair trade products are also organic. Consumers may link the attributes of these labels to each other, assuming that a fair trade product has been organically produced, or that an organic product encompasses fair trade practices.

US sales of fair trade coffee

According to Transfair USA, about 4.2 million pounds of coffee were fair trade certified in the US in the year 2000, 79% of which were also organic. In 2004, that number had risen to 32.8 million pounds, with 68% organic. (Remember that in 2000, federal standards for the organic label were not yet in place, so standards for organic coffee may have been varied). In 2003, the USA overtook the Netherlands as the largest destination for fair trade coffee, according to Transfair. US sales of fair trade certified coffee reached nearly $500m in 2005, with an estimate that at least 80% of that was also certified organic.

These dramatic increases are due in part to companies like Starbucks, Safeway, Nestlé and other large retailers responding to consumer and market pressure to include fair trade coffee options in their offerings. Even McDonalds agreed to sell, in a 650-store test market, fair trade and organic coffee sold by partners Newman's Own Organics and Green Mountain Coffee. *Nutrition Business Journal* writes, 'Starbucks does not disclose what percentage of the coffee dispensed at its 11,000 outlets is Fair Trade Certified, but it claims to be the largest purchaser of Fair Trade Certified coffee in North America. For the fiscal year ending October 2, 2005, Starbucks bought 11.5 million pounds, approximately 10% of global Fair Trade Certified coffee imports, according to a spokesperson. The company expected a small increase for 2006.'

Yet fair trade certified coffee accounted for about 4.5% of all specialty coffee sold in the US in 2005, and only 2.2% of all coffee sold. There is considerable opportunity for growth in this category. Transfair estimates that only about 15% of Americans have a high awareness of fair trade and its meaning.

Not everyone embraces the model that Transfair USA has created. As with organic, critics accuse the fair trade movement of having become too corporate and too bureaucratic, forcing farmers to pay for certification. *Reason* magazine published an article in its March 2006 issue titled, 'Absolution in Your Cup: The Real Meaning of Fair Trade Coffee,' (www.reason.com/0603/fe.kh.absolution.shtml). Author Kerry Howley writes, 'The movement has always aroused suspicion on the right, where free traders object to its price floors and anti-globalization rhetoric. Yet critics from the left are more vocal and more angry by half; they point to unhappy farmers, duped consumers, an entrenched Fair Trade bureaucracy, and a grassroots campaign gone corporate.'

Nonetheless, Howley writes, consumers like it, and fair trade sales continue to grow, not just in the coffee market but in other commodities as well. Neither the broader concept of 'ethical trade,' encompassing every link in the supply chain, or a domestic fair trade movement for US farmers has taken hold here yet, but the pathway toward these evolutions has been created by the fair trade coffee movement.

Peru was the top exporter of fair trade coffee into the USA in 2005, followed by Mexico and Guatemala.

Other fair trade commodities in the USA

Transfair USA gave fair trade certification to about 65 000 pounds of tea in 2001; by 2005, that had risen to 517 500 pounds, an increase of 187% over the previous year. They estimate that 87% of this volume was also organic.

Fair trade certified chocolate has also seen explosive growth. In 2002, Transfair USA certified about 14 000 pounds of cocoa; in 2005, they certified about 1.8 million pounds.

Transfair USA also certifies fresh fruit, including bananas, grapes, mangoes and pineapples, sugar and rice. These programmes are still in their infancy, but can surely be expected to capitalise on growing awareness of the fair trade label and its meaning, and to develop accordingly.

Consumer interest in other eco-labels

Despite many challenges, the organic model of clear standards and third-party certification has resulted in unmistakable success in the marketplace. By responding to attributes that consumers are willing to pay more for, and offering a guarantee of compliance, the third-party certification structure provides a basis for consumer confidence when 'buying local' is not an option. As a result, we're seeing many more agencies trying to develop similar labelling schemes.

Many consumers have said that they want more information about their food, especially in areas that the organic standards do not fully address. Among the plethora of 'eco-labels' appearing in the marketplace, however, some may confuse as well as clarify. Although some adhere to a highly meaningful and independently verifiable model, others may serve as marketing devices without a great deal of substance. If the organic label remains a gold standard of sorts, it remains to be seen which other labels will rise to the top; fair trade is certainly a well-established companion to organic.

In an interesting study of consumers' interest in food labels conducted by the Center for Agroecology & Sustainable Food Systems at the University of California at Santa Cruz, humane animal treatment ranked No. 1 among issues that respondents wanted clear labelling to support. This was followed by 'locally grown', 'living wage', 'US grown' and 'small-scale.'

Organic and eco-labels for non-food products

As the organic market and all that it implies expands, there is growing interest among both manufacturers and consumers for labelling criteria for non-food products. In the USA there has been controversy over standards for organic personal care products; greater clarity in the future for personal care standards can be anticipated. More definition in categories such as pet food, fish and dietary supplements is also likely.

The market for organic cotton and other 'eco-fashions' is also growing, after several years where it seemed to be somewhat stagnant. Wal-Mart has expressed great interest in organic cotton for clothing and textile production as well as in organic foods. Other major producers such as Patagonia and Nike continue to use organic cotton and small, entrepreneurial design companies are gaining visibility in the media and traction in the marketplace.

Conclusion

The organic movement has become a full-fledged and rapidly growing industry, with all of the rewards and challenges that that entails. With mainstream success comes a sense that vigilance will be required to keep integrity in the organic standards; the spectre of 'killing the goose that laid the golden egg' is definitely present. Meanwhile, interest in locally grown foods continues to develop.

The growth of the local foods movement is often positioned as 'local versus organic' rather than the two strategies being rightly seen as a distribution system and a production system that can complement one another in a food system that can truly approach sustainability. Many consumers are likely to adopt a purchasing philosophy that embraces both organic and local foods; they'll also look for other labels that resonate with personal values, such as humane animal treatment or small-scale farming. The result, ideally, will be a spectrum of choices that all contribute to a better and healthier world.

Appendix

Acronyms

These acronyms frequently appear in discussions of the US organic industry and marketplace.

AMS   Agricultural Marketing Service (a division of USDA)
CCOF   California Certified Organic Farmers (one of the first certifying agencies)
IOIA   Independent Organic Inspectors Association
NOP   National Organic Program (operates within AMS, a division of USDA)
NOSB   National Organic Standards Board
OCA   Organic Consumers Association
OFRF   Organic Farming Research Foundation
OTA   Organic Trade Association
TOC   The Organic Center
USDA   United States Department of Agriculture
EPA   Environmental Protection Agency
FDA   Food & Drug Administration

Trade, advocacy, research and regulatory organisations: contact information

Center for Food Safety
660 Pennsylvania Avenue SE #302
Washington, DC 20003
www.centerforfoodsafety.org

Consumers Union Eco-Label Project
101 Truman Avenue
Yonkers, NY 10703
914-378-2000
www.eco-labels.org

Environmental Working Group (EWG)
1436 U Street NW Ste 100
Washington, DC 20009
202-667-6982
www.ewg.org

International Federation of Organic Agriculture Movements (IFOAM)
Charles-de-Gaulle-Strasse 5
53113 Bonn, Germany
+49 (0) 228-926-50-10
www.ifoam.org

National Campaign for Sustainable Agriculture
PO Box 396
Pine Bush, NY 12566
845-361-5201
www.sustainableagriculture.net

National Organic Program at USDA
USDA-AMS-TMP-NOP
Room 4008-South Building
1400 Independence Avenue SW
Washington, DC 20250
202-720-3252
www.ams.usda.gov/nop

Northeast Organic Dairy Producers Alliance (NODPA)
c/o NOFA/VT
PO Box 697
Richmond, VT 05477
www.organicmilk.org

The Organic Center
www.organic-center.org

Organic Farming Research Foundation (OFRF)
PO Box 440
Santa Cruz, CA 95061
831-426-6606
www.ofrf.org

Organic Trade Association (OTA)
PO Box 547
Greenfield, MA 01301
413-774-7511
www.ota.com

Rodale Institute
611 Siegfriedale Road
Kutztown, PA 19530-9320
610-683-1400
www.rodaleinstitute.org

Rural Advancement Foundation International
PO Box 640
Pittsboro, NC 27312
919-542-1396
www.rafiusa.org

Transfair USA
1611 Telegraph Avenue, Suite 900
Oakland, CA 94612
510-663-5260
www.transfairusa.org

Companies: contact information

Aurora Organic Dairy
1401 Walnut Street
Boulder, CO 80302
720-564-6296
www.auroraorganic.com

Dagoba Organic Chocolate
PO Box 5330
Central Point, OR 97502
541-664-9030
www.dagobachocolate.com

Eden Foods
701 Tecumseh Road
Clinton, MI 49236
517-456-7424
www.edenfoods.com

Fairfield Farm Kitchens/Moosewood
309 Battles Street
Brockton, MA 02301
508-584-9300
www.fairfieldfarmkitchens.com

Horizon Organic
PO Box 17577
Boulder, CO 80308-7577
303-530-2711
www.horizonorganic.com

Newman's Own Organics
7010 Soquel Drive #200
PO Box 2098
Aptos, CA 95003
831-685-2866
www.newmansownorganics.com

Organic Valley Family of Farms/Organic Prairie
One Organic Way
LaFarge, WI 54639
608-625-2602
www.organicvalley.coop

Rudi's Organic Bakery
3640 Walnut Street
Boulder, CO 80301
303-447-0495
www.rudisbakery.com

Seeds of Change
3209 Richards Lane
PO Box 15700
Santa Fe, NM 87506
323-586-3455
www.seedsofchange.com

Small Planet Foods, Inc
1 General Mills Boulevard, 2BT
Minneapolis, MN 55426
763-764-7600
www.smallplanetfoods.com

Spectrum Organic Products, Inc
5341 Old Redwood Highway, Suite 400
Petaluma, CA 94954
707-778-8900
www.spectrumorganics.com

Stonyfield Farm
10 Burton Drive
Londonderry, NH 03053
603-437-4040
www.stonyfield.com

Straus Family Creamery
PO Box 768
Marshall, CA 94940
415-663-5464
www.strausmilk.com

White Wave Foods
1990 North 57th Court
Boulder, CO 80301
720-565-2344
www.silkissoy.com
www.whitewave.com

Whole Foods Market, Inc
601 North Lamar Suite 300
Austin, TX 78703
512-477-4455
www.wholefoodsmarket.com

Wholesome Harvest Organic Meats
PO Box 277
Colo, IA 50056
641-377-7777
www.wholesomeharvest.com

Wild Oats Market, Inc
3375 Mitchell Lane
Boulder, CO 80301
www.wildoats.com

Wild Sage Foods, Inc
16 Miller Avenue Suite 204
Mill Valley, CA 94941
415-388-7654
www.wildsagefoods.com

United Natural Foods, Inc
260 Lake Road
Dayville, CT 06241
860-779-2800
www.unfi.com

Veritable Vegetable
1100 Cesar Chavez Street
San Francisco, CA 94114
415-641-3500
www.veritablevegetable.com

Recommended periodicals

The Natural Foods Merchandiser (New Hope Natural Media)
1301 Pearl Street
Boulder, CO 80302
303-939-8440
www.naturalfoodsmerchandiser.com

Organic Processing
677 S.W. Tanglewood Circle
McMinnville, OR 97128
503-472-7387
www.organicprocessing.com

New Farm (On-line publication with organic price index)
The Rodale Institute
www.newfarm.org

Nutrition Business Journal (New Hope Natural Media)
www.nutritionbusinessjournal.com

Books

Lipson, E. (2001) *The Organic Food Sourcebook*. Mcgraw-Hill Contemporary, New York, NY.

Trade shows

Natural Products Expo West
Natural Products Expo East/Organic Products Expo/Biofach America

New Hope Natural Media
1301 Pearl Street
Boulder CO 80302
USA
866-458-4935
www.newhope.com
tradeshows@newhope.com

All Things Organic
Organic Trade Association
PO Box 547
Greenfield MA 01301
USA
413-774-7511
www.organicexpo.com

Chapter 13

Organic and Fair Trade Crossover and Convergence

John Bowes and David Croft

Setting the scene

In November 2005, market analysts Mintel produced a report (Organics UK) that claimed that organic food consumption in the UK had entered the mainstream by breaking through the £1bn sales barrier. Furthermore, the market researchers went on to forecast that organic foods would achieve sales of £2bn by the end of the decade.

The Soil Association, whose Annual Report had estimated organic sales for 2004 at £1.1bn, was even more bullish in response. They estimated sales could be as high as £3bn by 2010.[1]

This represents a truly remarkable performance. On the basis of these figures sales have effectively tripled in just five years. Organic foods, which held a modest 0.7% of the total UK food market at the start of the new millennium is now being forecast to have a market share of somewhere between 3% and 4% by the end of this decade.

This rate of growth, spectacular though it is, is dwarfed by the figures being estimated for the embryonic fair trade market. In 1999, UK food sales carrying the Fairtrade Mark stood at a modest £22m. By the close of 2004, they had increased almost sevenfold to annual sales of £140m (www.fairtrade.org.uk). While the organic foods market grew by 10% in 2004, fair trade, albeit from a smaller base, recorded sales growth in excess of 52% during the same period (see Figure 13.1).[2]

Patrick Holden, director of the Soil Association, said in response to the Mintel Report that 'There are very few people out there now who aren't worried about modern agriculture, climate change, food security, animal welfare, wildlife protection or the maintenance of public health. And those concerns are finding their expression in the market place.'[1] While this evaluation succinctly critiques the attraction of organic products, it offers no explanation for the phenomenal growth of the Fairtrade Mark, where the motivation for consumer behaviour must largely be driven by a humanitarian empathy for those suffering from the extremes of poverty in the developing world. It would be easy, therefore, but probably foolhardy, to believe

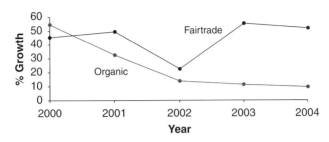

Figure 13.1 UK Fairtrade and organic sales.

that the success of these two markets was based on different groups of consumers driven by different functional and emotional needs. The success of these sectors is underpinned by changes in consumer attitude and behaviour, which reflect a widespread and almost instinctive recognition that all is not well with the world. These changes are tangible evidence of a desire to make a difference, however small the individual contribution, in halting a seemingly inevitable slide towards damage, catastrophe and disintegration in both a global and parochial context.

The ethical consumer

In 1994, CWS (the Co-operative Group) commissioned what is believed to be the largest survey of consumer attitudes to food and ethics ever undertaken. It showed that consumers were beginning to redefine their expectations from food as a result of concerns about issues such as animal welfare, the environment and third world poverty. They were increasingly seeking food that was sourced and marketed with integrity. The Co-op recognised this change in customer expectations and subsequently launched its ground breaking responsible retailing campaign in 1995.

Importantly, from an analytical point of view, the Co-op effectively repeated the initial survey a decade later, with a sample of almost 30 000 consumers. This second exercise, published in 2004, presents a unique insight to the changes in customer expectations during the preceding decade (Figure 13.2).[3] The results demonstrated that over the ten years, consumers had become even more concerned about the key ethical issues and more determined to seek out products with the right kind of credentials:

'Attitudes have hardened. Ten years ago, consumers were hungry for the ethical alternative and hungry for the information they needed to identify it. Now they are even hungrier. They are more prepared to use their purchasing power to support products that meet their standards – and equally prepared to veto those that don't. They increasingly want to play an active part in making these standards mainstream, using their influence to drive change.'[3]

	1994 %	2004 %	% Change
Are you more concerned now than in the past?	57	64	+12
Should retailers help growers in developing countries?	55	80	+45
Are you willing to pay a little extra for ethical alternatives?	62	84	+35
Have you boycotted a product on ethical grounds?	33	29	−12
Are you likely to boycott in the future?	60	60	0
Should food labels give full information?	62	96	+54
Should misleading labels be banned?	62	90	+56
Is it very important that retailers buy humanely-reared meat?	66	71	+7
Is it very important to support products not harmful to wildlife?	59	70	+18
Is it very important to stop products from non-sustainable sources?	55	64	+16
Is it very important that business minimise pollution?	52	67	+29
Is it very important that retailers minimise packaging?	52	58	+11
Average increase across all areas of the studies			**+23**

Figure 13.2   Co-op Ethical Survey. Source: Shopping with Attitude. The Co-operative Group, 2004.

What is quite startling about the results is not simply the increasing level of concern about ethical issues but also the overall magnitude of that concern. The interest is widespread with almost two-thirds of consumers being more concerned now than in the past. This might have been expected following a decade of food scares. Support for initiatives to help growers in developing countries, however, had increased by 45%, and embraced a staggering 80% of consumers surveyed in 2004. This suggests a growing level of altruism, coupled with a much greater expectation that large businesses should behave more responsibly, and be aware of the impact and influences that they may have.

Who cares about ethics?

Set in the context of this survey, both organic and fair trade food products are still, despite all of the impressive growth, performing well short of their true market potential. While both sectors are supported by consumers whose concerns may be specific, their overall growth and future growth potential seems to be based on a general upsurge in ethical awareness and concern that has a resonance with millions of consumers. Given the absolute scale of this interest, it is perhaps surprising that the performance of these products has not been even more impressive.

This dilemma is what Cowe and Williams refer to as the 30:3 syndrome, whereby a third of consumers claim to care about social responsibility but ethical products rarely achieve more than 3% market share. Their qualitative analysis identified five market segments:

(1) The '*Do What I Can*' group, which accounts for 49% of consumers, who while they have ethical concerns do not hold them particularly strongly and whose activism is muted.
(2) The '*Look After My Own*' group, accounting for 22% of consumers, who, as the designation implies, are the least concerned with ethical and environmental issues.
(3) The '*Conscientious Consumers*' who are active recyclers and generally interested in the issues while, at the same time, have no inherent feelings of guilt about consuming 'unethical' products. They account for about 18% of consumers.
(4) The '*Global Watchdogs*', about 5% of the total, who tend to be affluent, professional, middle-aged people, the ethical hardliners.
(5) The '*Brand Generation*', about 5% of consumers, who are young and brand conscious, but rarely exercise their spending power on ethical grounds.

Most of these groups express at least some concern about ethical issues but only the 'Global Watchdogs' pursue their interest in a consistent, active and committed way:

'The fierceness with which the hard core hold their beliefs, and pursue them in the shops is remarkable … They are, and feel, powerful, but want more information on which to base their decisions'.[4]

It is this group that underpins the current performance of both the organic and fair trade markets. They represent the key target for any immediate future growth but on their own do not represent large enough numbers to establish either sector as truly mainstream. The forward challenge will be to realise the full potential of the ethical hardliners while seeking opportunities to raise awareness, and break down the barriers to purchase that currently deter the generally interested, but altogether less active, customer groups.

In the beginning

While it is a fundamental change in consumer attitudes that has made the rapid development of organic and fair trade products possible, the development of both sectors was initially driven by a focus on producers rather than consumers. The organic sector in the UK didn't develop from a set of marketing concept boards but out of the genuine concern of a relatively small number of producers about contemporary farm practices. There was no overall master plan based on a perception of latent consumer demand, but recognition of the need to address the adverse impact that intensive farming techniques were having on the environment, animal welfare and local communities. This combined with an awareness – initially by just a few – that mass-produced food was gradually eliminating distinction and diversity and, as a consequence, this fuelled a determination to preserve food values through the production of high quality products in traditional ways.

Similarly, the initial development of fair trade products was not based on a conception of mainstream market potential but on the genuine concerns of a relatively small number of activists about the plight of disadvantaged producers in developing countries. These pioneers sought to establish a new trading model whereby producers achieved improved market access and received a fairer reward for their product through the mechanism of a guaranteed price and the payment of a social premium. They may have done this within an overall philosophy that embraced the need to eliminate poverty, through global trade reform, but they could have had little conception of how effectively their efforts might resonate with UK consumers. They set out to improve the lot of some of the poorest people on the planet rather than with the intention of satisfying the evolving needs of the ethically inclined consumer.

This is reinforced by the early marketing of these products, which focused on alternative trading organisations, such as Oxfam, which sold a limited number of products from third world producers. Sales were dependent upon a narrow but committed activist base. These early adherents to the fair trade cause were not just global watchdogs, they were global pioneers. As important and well intentioned as they were, however, a fundamental sea change was required if these products were to have any material impact in the marketplace. This came later with the emergence of fair trade organisations, development of certification and the Fairtrade Mark.

The underlying factors that have driven the development of both the organic and the fair trade symbols are many and complex, and not easily communicated to the end consumer. Education and a constant flow of information continue to play an important role in changing customer behaviour, but the absolute requirement was to find a simple mechanism which allowed consumers to make ethical choices without the need to become expert on complex environmental, social, financial and international political issues.

Both sectors have sought to address this issue through the development of universally recognised symbols, which offer the customer a simple and easily recognised guarantee of a product's authentic ethical credentials. In this respect, particular success has been achieved by the fair trade industry, which has largely been able to avoid the proliferation of marks, at least in any one individual country, which may have slightly hindered progress in the organic sector. The success of the Fairtrade Foundation's initiative can be judged from the knowledge that at least half of UK adults now recognise the Fairtrade Mark.[5] The development of this mark not only offered the customer a simple shorthand, implicitly communicating proof of audit and accreditation, it also opened up major new distribution opportunities by providing major retailers with a credible and recognisable vehicle through which to focus and deliver on their own ethical credentials. In addition, its resultant ubiquity helped to reinforce the fair trade message and the universal recognition of the central issues involved.

The evolution and development of the market from its activist base also presented key challenges in the context of product quality and retail price. While diehard loyalists might be prepared to pay premium prices for products that might even be perceived to be of lesser quality; the attraction of less committed customers required the right quality to satisfy a wider market and a careful balancing of the trade off between product quality and retail price. Given the inherent nature of the essential proposition, both organic and fair trade products were required to realise a price premium. Organics were also effectively marketed on a platform of superior quality, both in the sense of the intrinsic characteristics of the products themselves and the premium they delivered to the community through their method of production, but nevertheless, odd shaped fruit and vegetables were an issue with consumers whose expectations had been prejudiced by a retail industry dedicated to the merchandising of products whose key characteristics included visual perfection.

For fair trade the issues were a little different. Whatever the intrinsic quality of the basic raw materials some of the earlier products were not well profiled for UK tastes: for some, the initial eating experience created a lingering prejudice. Products achieving low levels of customer satisfaction were never likely to achieve significant market penetration, however strong their ethical credentials, while at the same time being marketed at a price premium. If fair trade was to breakout from its grassroots it needed to reposition its product offer on a much stronger quality platform.

Launch of the 'Cafédirect' brand of fair trade coffee and the 'Divine' chocolate brand from the Day Chocolate Company, represented pivotal moments in a shift towards the mainstream. Cafédirect helped to shape the standards by which the Fairtrade Mark would be defined. Today it is the fifth largest instant coffee brand in the UK. The Divine product was specifically profiled to UK tastes and positioned in direct competition to mainstream brands.

Making the breakthrough

While these developments were extremely important, the real breakthrough for fair trade came with key supermarket groups beginning to stock fair trade lines in mainstream distribution and the later development of 'own brand' fair trade products. In this context, the Co-operative Group, which effectively pioneered fair trade development between 1996 and 2003, acted as a major catalyst for the growth of the whole sector.

The Co-op was the first major retailer to stock CaféDirect, and in 2000, broke dramatic new ground when it became the first retailer to launch fair trade bananas, and introduced the first own brand fair trade chocolate. These were seminal initiatives. Bananas are not only the top selling fruit in the UK, but are also unquestionably the number one known value item (KVI). The introduction of the Fairtrade Mark to bananas shattered any doubts about the potential of fair trade to evolve from niche to mainstream.

The introduction of the Fairtrade Mark into the Co-op's own brand chocolate range was initially confined to a single product. This, in itself represented a minor initiative in terms of volume, but it also represented the very first stage in a development which would spearhead future growth. The incorporation of fair trade products into retailer own brand ranges has largely facilitated the repositioning of the 'fair trade brand' on a high quality, as well as an ethical platform. This platform, underwritten by the positioning of the retailers themselves, combined with their inherent volume potential, has generated accelerated performance improvement.

In 2002, encouraged by positive customer reaction and feedback, the Co-op extended fair trade to its entire own brand block chocolate range. This was followed in 2003 by the conversion of its entire coffee range to fair trade. Again, these represented seminal initiatives. Fair trade was no longer the niche product within the range: it was no longer a token gesture towards an ethical minority, nor carried the superficial gloss of corporate positioning; it was the standard by which others would be judged. Where one retailer bravely led, surely others would certainly follow.

The Co-op's commitment to fair trade reflected its unique origin and values. As an organisation it was committed to honesty, openness, social responsibility and caring for others. Marketing reflected this with a strong focus on high ethical standards through which it was able to protect its envious status as one of the most trusted brands on the high street. In fair trade the Co-op found a cause that matched its own values and facilitated its presentation to modern, forward thinking consumers, as the UK's most responsible retailer.

Raising awareness for the fair trade agenda has clearly been a key part of the Fairtrade Foundation's task. It has employed various methods to achieve this, perhaps most noticeably through the development of an annual Fairtrade Fortnight, which has received widespread support. With a committed retailer on board, seeking to drive fair trade as an important element in its own strategy, new opportunities to heighten levels of communication and raise overall awareness inevitably presented themselves. In the autumn of 2003, the Co-op produced a bespoke fair trade television commercial, as part of its award winning 'Creatures' campaign, which received heavyweight national exposure (Figure 13.3). It went on to adapt the commercial to support its main Christmas campaign: the fair trade message was transmitted

Figure 13.3   The Co-op's iconic 'Creatures' television campaign, December 2003.

simultaneously to millions of homes, on prime time television, during the most critical trading period in the year.

The organic sector also found a mainstream champion in the perhaps unlikely guise of Iceland (now the Big Food Group), a second tier retailer that made its name during the frozen food boom of the 1970s and 1980s. In June 2000, Iceland acquired around 40% of the world's organic vegetable crop with the declared intention of becoming an organic-only supermarket. Its strategy included a commitment to market organics at the same price as average supermarket own brand food, and the recognition that this would be achieved by a hefty cut in profits. This was part of a long-term investment to establish unprecedented ethical credentials.

Unfortunately Iceland, unlike the Co-op, was a listed company and therefore subject to the short-term disciplines of the City. When sales slumped the company was forced to refocus business back to its established trading platform. Large quantities of organic vegetables were offloaded into the marketplace and the first experiment in an organic-only supermarket was effectively over.

While this was disappointing, the reality was that a small retail chain like Iceland, however laudable its intentions, did not have not have the capacity to weather the initial difficulties, which were perhaps inevitable given the magnitude of the strategic change envisaged.

Waitrose, who started selling organic food as early as 1983, have had more sustained success by setting more realistic strategic objectives. Today they carry more than 1400 organic lines in their range, with 100% of their organic dairy products being procured from the UK. Organic foods are now believed to account for around 10% of Waitrose sales, far higher than any other UK supermarket. Nevertheless, as a relatively minor player in the UK market they do not have the potential, on their own, to deliver the levels of market penetration consistent with mainstream status.

The reality is that four retailers – Tesco, Asda, Sainsbury's and Morrison's – now dominate UK food retailing with almost 75% of the market. Accordingly, mainstream status cannot be achieved without the active involvement of these four players. Indeed, with Tesco's rapid development of the convenience sector, together with the remorseless growth of its core superstore business, it has not only strengthened its position as market leader, but also continues to outperform its competitors to such a degree that its future dominance of the market seems assured, unless there is a change in the regulatory regime. In short, it could be argued that Tesco's market penetration is now so great that both the organic and fair trade sectors must have its sustained commitment, or in all probability, suffer 'niche-dom' in perpetuity.

Fortunately, the Soil Association, deftly assisted by a series of high profile food scares, like BSE and foot and mouth, together with some powerful Royal patronage, has done an excellent job in bringing the retail giants on board. Sainsbury's made an early commitment to organics and by 2006 had built up their range to more than 900 lines. Perhaps more importantly, in November 2001, Tesco publicly announced its commitment to achieve organics sales in excess of £1bn within five years, representing 5% of their food sales. They backed this up by introducing hundreds of new products, as well as a commitment to support the Centre for Organic Agriculture at Newcastle University. Today, with well over 1000 products in their range, Tesco is clearly a dominant force in the UK organics market, as far as volume is concerned. Indeed, much of the recent impressive level of organics growth must be the result of their commitment to develop this sector.

Similar issues of expanding volume sales faced the Fairtrade Foundation. While support from empathetic organisations like the Co-op might be the bedrock upon which their initial success has been based, entering the mainstream demanded the participation of the major multiples and of Tesco in particular. While all four of the major multiples put an early toe in the market, the real breakthrough came in 2004 when Tesco introduced an own brand range of fair trade products, coincident with the tenth anniversary of Fairtrade Fortnight. By end of 2005, Tesco was stocking more than 90 fair trade lines and was able to claim that its customers bought one in three of the fair trade products sold in UK supermarkets. From virtually a standing start it had become the market leader for fair trade products in less than two years.

As with organics, it can be concluded that much of the recent growth in the fair trade sector has been driven by Tesco.

Dealing with the devil?

This success should be celebrated, yet it is not without its drawbacks. Retailers like Tesco are commercially driven. Their interest in developing the organic and fair trade sectors has almost certainly much more to do with anticipation of future customer demand than with any inherent ethical considerations. They are a large profit-orientated business with a constant eye on the customer as a vehicle for leveraging value for their shareholders. It follows that their commitment to these sectors is likely to be sustained for as long as they are in vogue and for not a minute longer.

Indeed, there may already be some evidence that Tesco is beginning to use its unquestionable muscle to try to influence change at the very heart of the fair trade concept:

'Tesco clearly views the expansion of its Fair Trade range as a success. However, a number of issues have arisen for the future. Tesco plans to expand its own-label offer ... but is constrained by the relative slowness of the Fairtrade Labelling Organizations' international certification process. Thus, current market opportunities in other categories such as fish, spirits and rice may well be missed. Furthermore, in order to expand the customer base beyond the traditional Fair Trade customer there would need to be some flexibility in the mechanics of the price floor for producers, to allow for lower-quality lower-priced Fair Trade goods (while still delivering to producers a fair price for the given quality).'[6]

Such views are likely to concern hard line activists who will inevitably view any movement away from the core concept as a compromise and for whom dealing with large multinational businesses (whose practices are perceived to represent the very issues that their campaigning is seeking to address) may represent the ultimate paradox and therefore be both rationally and emotionally unacceptable.

Some organic enthusiasts have also expressed exasperation at supermarkets who are attracted to the market by the high profit margins rather than ethics and whose inclination is to source their products from overseas rather than support the hard pressed local British producer. The *Organic Market Report 2005* by the Soil Association, while identifying continued high growth in organic sales, identified two points of concern. First, that the supermarket share of the market had declined from 81% to 75%. Second, an increase in the contribution made by imports, reflecting a switch away from UK-produced organic pork and beef.

At the heart of these issues is the fundamental distaste, which many activists and campaigners understandably have, for 'sleeping with the enemy' – they face a fundamental dilemma. Either they can remain true to the first principles of their ethical quest but remain niche participants in a drama they would prefer to direct, or they can sacrifice the purity of their vision, accept a position they perceive as second best, and potentially have a much greater material impact on the fundamental problems they originally set out to address. They can remain pure but ineffective or they can compromise and have a real impact.

Perhaps nowhere is this dilemma better illustrated than with the *bête noire* of the ethical movement. Nestlé have been the target of a boycott in no less than twenty different countries because, it is alleged, they aggressively market baby foods in the developing world, breaking World Health Organisation marketing requirements and, as a result, contribute to the suffering of millions of infants around the world. Accordingly, we can appreciate the dilemma faced by the Fairtrade Foundation when Nestlé communicated their intention to launch a variety of the Nescafé range that complied with the fair trade criteria.

Nestlé is one of the world's top five coffee roasters, who between them have been on the receiving end of a continuous campaign because of their role in pushing down raw material prices to a level so low that many small farmers have endured terrible suffering:

> 'There is a crisis destroying the livelihoods of 25 million coffee producers around the world. The price of coffee has fallen by almost 50 per cent in the past three years to a 30-year low. Long-term prospects are grim. Developing-country coffee farmers, mostly poor smallholders, now sell their coffee beans for much less than they cost to produce ... Farmers sell at a heavy loss while branded coffee sells at a hefty profit. The coffee crisis has become a development disaster whose impacts will be felt for a long time.
>
> Families dependent on the money generated by coffee are pulling their children, especially girls, out of school. They can no longer afford basic medicines, and are cutting back on food. Beyond farming families, coffee traders are going out of business. National economies are suffering and some banks are collapsing. Government funds are being squeezed dry, putting pressure on health and education and forcing governments into further debt.'[7]

In real terms, growers may be getting the lowest returns for a century, while the roasters have been posting record profits. Their financial results have been driven not simply by their huge buying power but also as major multinational companies, through their scope and ever increasing sophistication:

> 'Despite the stagnant consumer market, the coffee companies are laughing all the way to the bank. In the free market their global reach gives them unprecedented options. Today's standardised coffee blends may be a mix of coffees from as many as 20 different coffee types. Sophisticated risk management and

hedging allows the companies, at the click of a computer mouse, to buy from the lowest-cost producer to mix these blends'.[7]

These were issues that fair trade was designed to address and coffee has, arguably, been the Fairtrade Foundation's most successful venture, accounting for around a third of the total sales value for fair trade labelled products. As part of the campaign to address the imbalance in revenue distribution, and promote greater fairness and compassion, Oxfam called on the roasters to source at least two per cent of their beans from fair trade sources.[7]

Hence the enormous dilemma when Nestlé came forward with its plans for Nescafé Partners Blend. If this marked a fundamental change in position by one of the world's top roasters then it represented a great success and a massive breakthrough in the fundamental objective of achieving a fairer distribution of the spoils in the coffee market. If this represented the ethical awakening of a great multinational giant this development might represent the first in a series and lead the other roasters to consider their own strategies. If this represented a response to Oxfam's highly publicised campaign then how could it, in all conscience, be rejected whatever reservations the campaigners may have had about the motivation of the business involved?

That's a lot of 'ifs' and the reaction of many activists was both understandable and predictable:

'Unison, the public service union, branded the move 'cynical' while War on Want, a charity dedicated to eradicating global poverty, complained that the Fairtrade Foundation logo was set up to challenge the dominance of multinational companies such as Nestlé. Others said Nestlé's involvement demeans the Fair Trade logo'.[8]

In reality the activists could only carry the fair trade argument so far. Without the development of the Fairtrade Mark and the involvement of the supermarkets they couldn't really expect to have much impact. Having once taken the irretrievable step of involving the big retailers in their campaign there was simply no going back. Their purest roots were abandoned, for the very best of reasons, with the fair trade endorsement of own brand coffee, wine and chocolate. Getting into bed with Nestlé was just the next inevitable step.

Time will tell whether Nescafé Partners Blend represents a cynical absorption of the Fair Trade campaign, or the first stage of fundamental repositioning of the Nestlé brand, with a potential self-evident impact on the rest of the market. From the campaigners' perspective they might be reassured by the realisation that Nestlé, still suffering assaults from baby milk campaigners, would not easily want to court another public relations disaster. In the long run it is inevitable that Nestlé, like Tesco, will be driven by the pecuniary interests of its shareholders. Conversion to a more ethical way of doing business is likely to depend on customer reaction. If

Nescafé Partners Blend receives overwhelming consumer endorsement then it could set the standard for the whole industry.

Whatever the risks, the Fairtrade Foundation had little choice but to endorse and support the Nestlé move. Similarly, the Soil Association, would naturally, and pragmatically, seek to address any concerns about supermarkets' retail pricing through a process of dialogue and diplomacy in preference to outright confrontation:

> 'A study authored by Dr Anna Ross (University of the West of England: www.corporatewatch.org.uk/?lid=252) accuses UK supermarket chains of overpricing organic goods. Media reports suggest that the Soil Association has been trying to suppress the findings. Dr Ross found the same basket of vegetables bought in a sample of farm shops were found to be 63% more expensive in market leader Tesco; 59% more expensive in Sainsbury's; and 38% more expensive in Waitrose. On average, she found the cost to be 64% more expensive in the multiples than local farmers' markets. Dr Ross accused the Soil Association of being too busy trying not to upset the supermarkets, and encouraged consumers to shop elsewhere for better value.'

These kinds of problems reflect the success of campaigning groups. They have gained a very high level of awareness of the key issues, and stimulated a level of consumer interest sufficient to persuade major businesses to reconsider their current strategies. It is inevitable, as part of that process, they will face uncomfortable decisions and, if real success is to be achieved, may have some of the fundamental tenets of their own strategies challenged.

New challenges ahead

Organic farmers aim to produce good food from a balanced living soil using crop rotations to make the land more fertile. The use of artificial chemical fertilisers and pesticides are severely restricted and farm animals are reared without the routine use of drugs and antibiotics, which are common in intensive livestock farming. The aim is to produce food that is safe, healthy for consumers, less damaging to the environment, more natural for the animals and that ultimately tastes better. As a result the organic symbol represents a 'gold standard' but it may not represent the most effective route to Valhalla.

An alternative cropping strategy through Integrated Crop Management (ICM) seeks to conserve the environment while economically producing safe and wholesome food. It attempts to minimise the reliance on fertilisers and crop protection chemicals but without the same degree of restriction, legally enshrined in the Organic Standard. As a result, its use as a technique falls short of the 'gold standard' but involves considerably less risk, and delivers more certain yields, while significantly reducing the adverse impacts associated with intensive farming.

While the Organic Standard delivers the greatest benefit, it only achieves this with greater risk to the producer, only partially offset by the promise of greater gains paid through a substantial premium by the consumer, a premium that is offputting for many. In addition, despite the impressive growth of the sector in recent years, it is still expected, even on the most bullish projections, to take only a modest share of the UK food market. ICM, on the other hand, involving less risk, higher yields and lower costs of production, might be capable of achieving rapid penetration within UK agriculture with the support of the major farm groups, if an effective means of marketing its benefits to consumers could be found. The scale and scope of ICM's potential may be sufficient to dwarf any additional measurable benefit, such as a reduction in pesticide use, which might be achieved in the foreseeable future through a purest adherence to organic principles.

One alternative for campaigners might be to adopt appropriate ICM protocols, as a 'silver standard', receiving formal endorsement and support, while continuing to pursue the Organic Standard as the long-term end game. Such an endorsement might represent a bitter pill to swallow for many activists, but the benefit might be very rapid, sizeable, and deliver demonstrable improvement in UK farming practice. Indeed, the alternative may prove less palatable as UK retailers, always with an eye to the market, recognise the opportunity and use their own considerable brand power to exploit its consumer potential.

Tesco introduced 'Nature's Choice' as its own integrated farm management scheme in 1992. The standards include a commitment to the rational use of plant protection products, fertilisers and manures and it applies to all fruit, vegetable and salad suppliers to UK stores. However, the initiative has drawn criticism from activists and campaigners:

> 'Nature's Choice' supposedly encourages 'rational' pesticide use, but details of the scheme are not publicly available. Tesco says it works with suppliers to keep pesticide use to the minimum required, yet it refused to sign a Friends of the Earth (FOE) pledge to take action to deal with risky chemicals ... Despite its claims, FOE's analysis of government data on Tesco for five years from 1998 to 2002 showed that Tesco had made no overall reduction in pesticide residues in its food. Over the five years, an average of 45% of Tesco fruit and vegetable samples contained pesticide residues'. (www.corporatewatch.org.uk/?lid=252)

Such controversy, justified or otherwise, might perhaps have been avoided, and more widespread commitment to environmental improvement been realised, if the Soil Association, or any other appropriate NGO, had taken the initiative to establish an appropriate certification standard for ICM, and developed a universally recognised mark.

The fair trade sector faces similar challenges. The very rapid growth of the sector has created its own stresses and strains. Sophisticated modern retailers with strong own brand ranges expect to be able to move from initial product concept to on shelf delivery with great speed. While the fair trade movement has been very successful in creating high interest in its potential, it has, with obviously limited resources, not always been able to keep pace with the development expectations. The launch of the Co-op's Carmanere Chilean wine, the first ethically traded wine in the UK, was done in collaboration with Traidcraft, and did not initially carry the Fairtrade Mark because the appropriate criteria were not at that time, in place. More recently, Tesco is believed to have expressed concern at the slowness of the whole certification process.[6]

Such tensions can potentially create problems, which, if not handled with the greatest of care, could undermine the credibility of the whole fair trade programme:

'Critics of a Fairtrade growth model that relies too much on market information and high-growth partners worry that short-term market growth strategies are sacrificing the long-term viability of the Fairtrade system. First, growing suppliers and traders faster than the certification system can realistically audit producers and supply chains exposes the certification system and its independent guarantee to failure and possible scandal, damaging the integrity of the Fairtrade mark and weakening the entire system. Similarly, rushing through standards development for new Fairtrade products risks having price floors or trading standards that are not fully researched and thus do not truly ensure the sustainable development and fair prices that consumers associate with the Fairtrade label.'[6]

This represents a real challenge to the Fairtrade Foundation. True success can only be achieved through collaboration with the most vibrant and successful businesses acting in response to customer demand, but the fair trade model, with its coherent, disciplined, and fundamentally necessary audit procedure, by its very nature, represents a bottleneck that frustrates partners, slows the realisation of market potential and, in consequence, the flow of benefits to disadvantaged growers in the developing world.

There is no easy solution to this issue. The obvious danger is that the largest and most successful retail partners, frustrated by the process and alive to profit opportunity, may elect to plough their own furrow. Their brand power is so great that, unless the situation is managed very carefully, the opportunity to develop 'Fair Choice' or 'Fair Reward' or some such sub-brand mechanic, within their own brand ranges, with less restrictive criteria and controls, may become a real possibility.

This potential proliferation of marks may not be limited to big brand retailers. The Rainforest Alliance is an international conservation organisation based in the United States. Its declared mission is to protect ecosystems, the people and wildlife that live in them, by encouraging better business practice. For growers to be certified by the alliance they must adhere to sustainable agricultural principles that include preserving local wildlife, protecting forests, minimising soil erosion and treating workers fairly. The emphasis is on protecting the environment while at the same time giving growers the opportunity of a better life through entry into premium coffee trading. While there are undoubtedly genuine benefits in this approach, critically there is no guaranteed price to the growers or any formal social premium, yet Kraft Foods have recently chosen to support the Rainforest Alliance to launch their ethical offers, rather than the Fairtrade Foundation:

'So what is the attraction to food companies of the alliance's scheme? Cynics will point to the price. Not only is Rainforest Alliance certified coffee cheaper than the fair trade alternative, there is also no licensing fee to use the alliance's logo. The Fairtrade Foundation, on the other hand, charges a 2% fee, based on the wholesale coffee price – another significant cost. This all makes the Rainforest Alliance a cheaper way for the large coffee brands to tap into the ethical market. Coffee roasters that sell coffee containing a minimum of 30% certified coffee beans can boost their ethical credentials by using the Rainforest Alliance logo on packaging.'[9]

The clear implication is that the Rainforest Alliance's real emphasis is much more on environmental standards than on delivering fairer trade. While the fair trade criteria require producers to implement environmental improvement plans, the emphasis is much more upon the essentially social goal of creating the opportunity for growers in the developing world to gain a greater control over their own lives. The emphasis is much less on environmental management and much more on working conditions and fair prices. It is true that many of the producers use organic methods but this largely reflects the existing practices of many small third world producers rather than a fundamental tenet of the fair trade criteria.

Convergence and divergence

Nevertheless, this apparent, if fragile, convergence of both social and environmental goals has resulted in a close association of both the fair trade and organic sectors in the UK. The basic standard of the International Federation of Organic Agriculture Movements (IFOAM) embraces social guidelines and stipulates the requirement for workers to have decent conditions of employment. In this context, the Soil Association has long argued that it is not just growers in the developing world that receive insufficient remuneration to cover their production costs, but also many hard pressed British farmers. Accordingly, in the pursuit of potential synergies the Association, working initially with the Fairtrade Foundation, has developed a pilot ethical trading scheme.

The emphasis is on ensuring a fairer return for farmers, and the fair treatment of workers, while making a positive contribution to the local community. The criteria for the scheme are positioned as additional and voluntary standards, to complement the existing organic standard. The process of certification addresses areas such as access for workers to trades unions, health and safety, transparency and accountability, fair distribution of risks and rewards, composting and recycling, as well as local marketing, and assesses the whole of the supply chain before awarding the right to use the ethical trading symbol.

However, few true synergies with the fair trade sector emerged from the co-operation between the two key associations. The Soil Association would probably have preferred to complement their organic standards with a Fairtrade Mark. The Fairtrade Foundation have argued, not unreasonably however, that the Fairtrade Mark, and indeed the very words 'fair trade' themselves, have become firmly associated in the mind of the public with the plight of the real poor in the third world. However difficult the problems faced by small organic farmers in the UK, they are at least protected by a social, economic and political infrastructure that ultimately protects them from the full force of an adverse market. Their children will not go short of an education, they have access to a free health system and they are ultimately protected by a social security system that should ensure that they will not starve.

The whole concept of fair trade has captured the public imagination in a quite extraordinary way. The evidence for this is not simply from the growth in product sales, or the widespread recognition of the Fairtrade Mark, but from the almost ubiquitous recognition and understanding of the term itself and an extensive grass-roots endorsement through developments such as fair trade towns.

There are, at time of writing, more than 230 towns in the UK that have successfully gained fair trade status by securing the active support of their local council and achieving widespread availability of fair trade products in local shops and workplaces. This is probably an unparalleled achievement. Clearly, it owes a lot to local activists who have been driven by a humanitarian agenda: it could not have been achieved without the almost universal endorsement of their local communities. For millions of people, fair trade has become a simple pain-free mechanism for expressing positive empathy with and support for the third world poor.

In this, fair trade differs sharply with the organic movement. Whatever similarities there may be in their equally Herculean tasks, whatever the crossover of issues and the convergence of tasks, when it comes to understanding the motivation of customer support, the two sectors are poles apart.

The purchase of organic products is primarily driven by the expectation of personal benefit. The customer anticipates that the product will both taste better and be a healthier alternative to the mass produced, chemically enriched, pesticide-soaked products which, in the twenty-first century, seem to have become the standard fare. This does not mean that the customer is not ethically inspired and does not feel better for choosing to consume a product that is less damaging to the environment. Purchasing organic products delivers both functional and emotional benefits. Arguably, though, the primary or functional motivation of the organic consumer is to purchase a better and healthier product.

For fair trade the motivation is very different. There is no expectation that the product itself is likely to be intrinsically better or healthier, and yet the consumer has to pay a premium for its purchase. It has no functional benefit in comparison with less ethical competitive offerings. In reality its functional and emotional benefits are fused. The fair trade consumer feels positive about making a simple and inexpensive contribution to the less advantaged in the developing world. The primary motivation for purchase is to benefit the world's poor.

The validity of this hypothesis is perhaps demonstrated by the relative weakness of fair trade in the United States where the prevailing culture of self-centred political and emotional isolationism, together with an undeviating commitment to the capitalist market economy, drives the single-minded pursuit of personal material wealth at the expense of an empathetic response to global economic and social problems.

While altruism might set fair trade apart, it also represents a potential threat to the future of the sector. If the primary function, from a customer perspective, is to affect wealth transference from the rich to the poor, then fair trade, assessed in the context of its efficiency, certainly has some questions to answer. Any classically based economic analysis of its role is bound to question its effectiveness in comparison with the long-term benefits likely to flow from free trade. Alex Singleton, writing in *The Business*, described it as 'a combination of economic illiteracy and do-gooder foolishness', which 'had created a monster that threatens the prosperity of the poorest producers'.[10] Whether or not we subscribe to the economics of Adam Smith, there is plenty of evidence to suggest that, as a tool for wealth transference, fair trade seems to represent a particularly inefficient vehicle. Concerns have already been expressed at the relatively small amount of benefits that the fair trade premium yields to the growers. John McCabe, writing about fair trade banana pricing, said that consumers 'will be surprised to learn that the lion's share of the extra money appears in many cases to be going to the supermarkets. Much of the premium over and above the fixed price they must pay the farmers appears to be growing straight into their profits'.[11] These comments reflected an analysis that concluded that while farmers were earning approximately 24 pence on a kilogram of fair trade bananas, the supermarkets were pocketing between 35 and 65 pence dependent upon their retail pricing. Similar criticisms have been made about fair trade coffee pricing:

> 'But the most dismal return found was between Nescafé and Cafédirect, a Fair Trade brand found in many UK supermarkets. Mendoza and Bastiaensen's (2003) comparison of returns to Nicaraguan coffee farmers through Fair Trade versus conventional marketing chains found that in 1996 with a high-priced coffee market consumers paid $1.63 (£0.91) per pound extra for Cafédirect over Nescafé, but the Nicaraguan coffee farmer saw only $0.03 (£0.02) extra, a return of just 2 per cent'.[6]

While the precise benefit the fair trade grower receives may fluctuate in response to the movement in world market prices, and the policies of different actors in the supply chain, it is difficult to contest the assertion that the essential mechanism itself distorts the level of benefit that most consumers would anticipate would be returned direct to the intended beneficiaries.

Apart from the obvious costs associated with the management and administration of the scheme, fair trade products also incur heavy supply chain costs including high retailer margins. As a result, only a small percentage of the price paid by the customer makes its way back to the developing world. Accordingly, if the customer's intention when making a fair trade purchase is primarily to pass a benefit to the poor and needy then it would make far more sense for them to make a simple direct charitable contribution.

This assessment of fair trade as a complex and rather inefficient cause-related marketing tool is understandably and fiercely resisted by the sector's activists. Their argument is that it is not charity but a change in the trading process driven by ethically inspired consumers. The process delivers an obvious benefit in terms of better prices for the growers but also additional benefits arising from the payment of the social premium to support development projects such as the building of schools or delivering better access to clean water. Fair trade also supports the development of co-operatives, which provide the infrastructure to help sustain development as well as facilitating the direct participation of small growers in determining their own futures. Some of these benefits may be difficult to measure but they are likely to be very important in establishing two key factors for the growers' sustainability – confidence and self-belief.

Managing momentum

The success of fair trade, whether viewed as trade or aid, demonstrates the validity of the results from the Co-op's 'Shopping With Attitude' survey.[3] In reality, there is an enormous empathy in the UK with third world poverty. Television images projected regularly into millions of homes combined with massive publicity events, such as Live Aid, have stirred the conscience of millions of people. Many of them buy fair trade products because they want to feel they are making a little difference without, in any serious or material way, affecting their own rather comfortable lives. In purchasing the product they believe themselves to be making an ethical choice.

The development of the ethical consumer does not simply impact on individual product choice. The problems Nestlé have encountered demonstrate the difficulties that businesses may face if they are not alive to changing customer expectations. Campaigning groups and an ethically savvy media can, in an instant, draw consumer attention to discrepancies from expected standards, unleash sustained public criticism and damage corporate reputation. As a result, many companies have developed formal policies on corporate and social responsibility in order to protect their reputation and their brand.[12]

This development has important implications for both the fair trade and organic sectors. Companies, particularly food retailers, wishing to demonstrate their corporate and social responsibility, recognise the need to show empathy with customers' changing expectations about the developing world, the environment, food integrity and local products. The existence of recognisable label marks for both the organic and fair trade sectors paradoxically reduces the immediacy of the challenge by presenting businesses with a simple and inexpensive means to demonstrate ethical compliance. A recent report from UNEP concluded that eco-labels represented an easy and effective way for major companies to display ethical credentials to consumers, government agencies, investors and employees.

While in the short term businesses may succeed in demonstrating their 'good-guy' credentials by stocking fair trade and organic products in their stores, canteens and vending machines, this is unlikely to represent a successful model for demonstrating corporate and social responsibility in the future. The essential issues are of scope and scale. Consumers and the media will ultimately demand much more than token ethics. Their expectations will search out and demand a much more holistic commitment to corporate and social responsibility.

The problem for both the Fairtrade Foundation and the Soil Association is that their current schemes cannot match that expectation. The nature and requirement of the current certification and monitoring procedures will limit the extension and growth of the potential for fair trade. The singular focus on the organic gold standard will ensure that the Soil Association will remain an important, rather than a major, influence on UK agriculture.

The likely outcome is that major companies will increasingly fall back on their own resources to demonstrate their global citizenship. Thus, for example, the development of ethical initiatives focusing upon compliance with defined workplace standards, rather than upon a redefinition of the supply chain is likely to become more prevalent. Such initiatives will require a much more collaborative approach between parties within the existing supply chain. The emphasis will be upon suppliers to generate solutions and ensure compliance.

The absence of any formal certification procedure from such schemes will, initially at least, be seen as a fundamental drawback. The sheer scale of the task, however, probably makes this inevitable in order for the greatest progress to be made. Attempting to achieve compliance through inspection and certification of the tens of thousands of primary producers involved in supplying UK supermarkets, would be simply impossible. The alternative would be to create and rely on a system that might only bring benefit to a modest few.

These kinds of issues represent a major challenge to the fair trade and organic sectors, but the fact that they may be on the horizon at all is a testament to the tremendous progress they have made in recent years. From humble beginnings they have campaigned to forge a real niche in the UK food market. Both conceived at the instigation of a few activists, they have exploited a major change in customers' attitudes to deliver both sectors to the brink of mainstream status. The obstacles they have faced are huge. The similarities of the challenge, and the proximity of many of the issues, suggest a great commonality of cause, but in reality their goals are very different. While the organic sector appeals strongly to the most ethical consumers, its success is primarily based on enlightened self-interest. In contrast, the success of the fair trade sector essentially represents the commercial manifestation of an altruistic expression of human sympathy.

Paradoxically, success will now generate their greatest challenge. The break into mainstream will necessitate continued support from the 'global watchdogs' as well as the conversion of the 'conscientious consumer'. This task can only be achieved by engaging with the major multinational brands irrespective of the level of angst this may cause for the most purist campaigners. The potential downsides are many, but even the most optimistic outcomes present clear and difficult challenges. The truly engaged big beasts, alive to rapidly changing customer expectations, may want to move further and faster than the sectors can cope. Possessed of enormous brand power, and a perceived need to extend the scope of their activities on a more holistic basis, they may conceive that greater progress might be made by seeking their own bespoke solutions in order to underpin a corporate requirement to demonstrate social responsibility.

Potentially this presents a major dilemma for the key NGOs. On the one hand, they can stand firm against the tide. They can remain comfortable with the constant self assurance that their conception of the challenge represents the only true solution. They can be evangelical, consistent and increasingly less relevant.

Alternatively, they could anticipate the changing situation ahead, remaining committed to the essential key needs which they exist to address, and seek to develop their strategies in order to maintain their influence and relevance. They can adapt, survive, and help change the world.

References

1 Hickman, M. (2005) Sales of organic food and drink to reach £2bn by 2010. *The Independent*, 28 December.
2 Co-operative Bank (2005) The Ethical Consumerism Report 2005. www.co-operative-bank.co.uk
3 Co-operative Group (2004) Shopping with Attitude. www.co-op.co.uk.
4 Cowe, R. and Williams, S. (2000) *Who Are the Ethical Consumers?* Co-operative Bank/ MORI, Manchester.
5 Fairtrade Foundation/MORI (2005) Awareness of the Fairtrade Mark rockets to 50%. http://www.fairtrade.org.uk/pr270505.htm
6 Nicholls, A. and Opal, C. (2005) *Fair Trade*. Sage Publications, London.
7 Oxfam GB (2002) Mugged: Poverty in your coffee cup. www.oxfam.org.uk
8 Hall, J. (2005) Has Nestlé spiked Fair Trade's coffee? *Sunday Telegraph* (Business), 30 October.
9 McAllister, S. (2004) Who is the fairest of them all? *The Guardian*, 24 November.

10 Singleton, A. (2006) Why 'fair' trade is a bad deal for poorest farmers. *The Business*, 12 March.
11 McCabe, J. (2003) *The Sunday Times*, 29 June.
12 Barrientos, S. and Dolan, C. (Eds) (2006) *Ethical Sourcing in the Global Food System*. Earthscan, London.
13 Rotherham, T., Associate, International Institute for Sustainable Development (IISD) (2005) *The Trade and Environmental Effects of Ecolabels: Assessment and Response*. UNEP, 25 November.
14 Barrientos, S. and Dolan, C. (Eds) (2006) *Ethical Sourcing in the Global Food System*. Earthscan, London.

Further Information – Useful Organic and Fair Trade Websites

Simon Wright

Putting the words 'organic food' into Google generates 83 million weblinks: the word 'Fairtrade' generates a mere 3.5 million hits. With so much information available some guidance is needed as to where to begin looking for more information about matters organic and fair trade. The websites below are recommended as useful sources of further information in this field. US websites are listed at the end of Chapter 12.

Organic brands

www.abel-cole.co.uk
www.amys.com
www.aspall.co.uk
www.clipper-teas.com
www.communityfoods.co.uk/crazyjack.html
www.dovesfarm.co.uk
www.duchyoriginals.com
www.fetzer.com
www.freshnaturallyorganic.co.uk
www.graigfarm.co.uk
www.greenandblacks.com
www.grovefresh.co.uk
www.helenbrowningorganics.co.uk
www.hipp.co.uk
www.junipergreen.org
www.lymeregisfoods.com
www.naturespath.com
www.organix.com
www.rachelsorganic.co.uk
www.rdaorganic.com

www.seedsofchange.co.uk
www.sheepdrove.com
www.shipton-mill.com
www.smallplanetfoods.com
www.uk5.org
www.village-bakery.com
www.wholeearthfoods.com
www.villagebakery.co.uk
www.yeovalleyorganic.co.uk

Organic organisations

www.agroeco.nl
www.biobank.it
www.condor-organic.org
www.cremoniniconsulting.com
www.defra.gov.uk/farm/organic
www.ec.europa.eu/comm/agriculture/qual/organic/
www.efrc.com
www.fao.org/organicag
www.fibl.net
www.food.gov.uk/foodindustry/farmingfood/organicfood
www.gardenorganic.org.uk
www.greenplanet.net
www.grolink.se
www.ifoam.org
www.omri.org
www.organic.aber.ac.uk
www.organicandfair.com
www.organicfqhresearch.org
www.qlif.org
www.soilassociation.org

Organic certifying bodies

www.australianorganic.com.au
www.biodynamic.org.uk
www.bioland.de
www.ecocert.com
www.ics-intl.com
www.nasaa.com.au

www.natureetprogres.org
www.naturland.de
www.organicfarmers.org.uk
www.orgfoodfed.com
www.qai-inc.com
www.skal.com
www.soilassociation.org/certification

Fair trade brands

www.agrofair.com
www.cafedirect.co.uk
www.clipper-teas.com
www.divinechocolate.com
www.dubble.co.uk
www.equalexchange.co.uk
www.traidcraft.co.uk
www.tropicalwholefoods.com

Fair trade organisations

www.fairtrade.net
www.fairtrade.org.uk
www.organicandfair.com
www.twin.org.uk

Retailers

www.alnatura.de
www.basicbio.de
www.co-opfairtrade.co.uk
www.sainsburys.co.uk/food/foodandfeatures/sainsburysandfood/ourranges/
Organics+range.htm
www.sainsburys.co.uk/food/foodandfeatures/sainsburysandfood/fairtrade
www.tesco.com/health/eating/?page=organic
www.waitrose.com/food_drink/wfi/foodissues/organicfood.asp
www.waitrose.com/search.asp?q=fairtrade&x=3&y=7

Consumer orientated

www.nicecupofteaandasitdown.com
www.observerfoodmonthly.co.uk
www.organicconsumers.org
www.organic-life.net
www.whyorganic.org

Trade magazines and information services

www.biohandel-online.de
www.dusolalatable.com
www.ekoconnect.org
www.ekolantbruk.se
www.ekoweb.nu
www.naturalproducts.co.uk
www.naturasi.it
www.newhope.com
www.oekoinform.de
www.linksorganic.com
www.organic-business.com
www.organicdenmark.com
www.organic-europe.net
www.organic-market.info
www.organicmonitor.com
www.organicstandard.com
www.soilassociation.org/web/sacert/sourcemarketplace.nsf

Trade shows

www.biofach.de
www.natexpo.com
www.naturalproducts.co.uk/np%5Feurope
www.vidasana.org

Index